V188
My Torpedo Boat War Missions

The Great War at Sea Series

Volumes in the Great War at Sea Series

V188 My Torpedo Boat War Missions. Friedrich von Callisen

Front cover art: Willy Stöwer's 1917 depiction of V44 launching torpedoes at the battle of Jutland.

V188
My Torpedo Boat War Missions

Friedrich von Callisen
Lieutenant and Commander

Published by Trident Publishing
Miami, Fl USA

Library of Congress Cataloging-in-Publication Data

Callisen, Friedrich von
Originally published as V188 Mein Torpedoboot Kriegsfahrten, Berlin 1916
My TorpedoBoat War Missions
ISBN-13: 978-0-9963157-0-8
World War, 1914-1918 -- Personal narratives, German.
World War, 1914-1918 -- Naval operations.
Military operations, Naval.

Printed in the United States of America

Translator's Foreword

The centennial of the Great War has produced a bumper crop of both republished and newly discovered memoirs and secondary works relating to that conflict. British Commonwealth and Empire accounts predominate among English language publications as is to be expected. It is appropriate to conclude that readers and researchers would benefit from examining accounts from the "other side of the hill"[1] and examine German accounts for corresponding actions and sentiment in like minded endeavors.

Friedrich von Callisen's wartime memoir is remarkable on many levels: 1. It is the only memoir of a German torpedoboat commander comprising the light forces which, next to submarines, saw the majority of combat and suffered disproportionate casualties.[2] 2. He experienced and survived considerable combat and vessel sinkings. 3. The work was created during the war yet manages to escape the pallor of wartime propagandizing, fictionalization, security controls and other attributes which degrade war era purported non-fiction to fiction.[3] 4. Although

---

[1] B.H. Liddell Hart coined the phrase in his work examining German senior officer accounts of the Second World War. Of course, the expression dates to an anecdote regarding the Duke of Wellington.

[2] 56 German torpedoboats and larger destroyers were sunk due to mines; 18 by surface gunfire, 3 by submarine torpedoes, and 18 due to operational reasons as in grounding, collision, etc.

[3] For example, the majority of wartime German submarine commander

brief, it is well regarded in its position as part of German Great War literature.[4]

A brief examination of the German torpedoboat service demonstrates its personnel were specialized and an elite within an organization which promoted the concept of eliteness. Officers specializing in torpedoes trained extensively when assigned to torpedoboat duty to maintain their "T" as the certification was informally known. They also served periods on larger warships to increase breath of experience as the rigors of torpedoboat duty were primarily for younger officers.[5] Warrant, Non-Commissioned and Enlisted personnel were genuine torpedoboat career specialists and, at least in the prewar period, the pick of men assigned to the navy. Once in the torpedo service a man stayed in the service unless a delicto forced their ejection.[6]

---

printed accounts are notorious for fabrications. Callisen's account matches entries in his 1st Torpedoboat Flotilla War Diary and is confirmed by other sources footnoted in this text which include the German Official History and eyewitness reports by adversaries.

[4] The Bundesarchive summary of holdings examines German wartime literature and indicates Callisen's work is the sole example where a commander mourns the loss among his crew after incurring combat casualties rather than engage in jingoistic patriotic fervor.

[5] A torpedoboat command was typically a Sr. Lieutenants billet. Callisen being about 29 years of age when assigned to command V188. A Flotilla commander being a Korvettenkapitan or Lieutenant-Commander in USN parlance. Appendix A details his career.

[6] The USN attaché in Berlin, Reginald Belknap, reported torpedo boat men wear a distinct identifier on their cap. The reference is most likely due to the carmine piping on the cap tally used by enlisted torpedoboat and submarine sailors whereas the rest of the navy did not. Note in the relevant Appendix Callisen's report regarding the Loss of V188 appended after the text the Other Ranks are prefixed with "Torpedo" next to their specialization onboard.

There were eight,(nos. 1-8), Torpedo Boat Flotillas in the Order of Battle at the outbreak of war with another added in December 1914.[7] Flotillas were generally grouped into two half-flotillas of five boats plus one additional leader boat to complete the organization of eleven boats per flotilla.[8]V188 was part of the 1st Half Flotilla of the 1st Torpedo Boat Flotilla home ported in Kiel.

Torpedo boats were built by three builders to maintain design homogeneity within flotillas and half-flotillas- Schichau, whose boats have an "S" prefix, Vulcan, a "V" and Kiel whose boats bear a "G". Torpedoboats were numbered not named. V-188 completed in 1911 of about 666/775tons, within a slim design of 73.9m length x 7.56m breadth, housing four coal fired boilers generating 18000 horsepower and up to 32 knots and a crew of 84.[9] The vessels were designed for torpedo attack on larger warships and minimized own artillery, only two 88mm artillery pieces were fitted, and relied on four 500mm diameter torpedoes in above water tubes as their main weapon. Flotilla sister boats are of a similar design.[10]

The 1st Torpedo Boat Flotilla was nearing the end of a three year commission in 1914. Callisen was to be the only commander of V188. The crew would have been

---

[7] Two more plus a Flanders Flotilla were formed during the war. Does not include second class units used as harbor defence, escort duties, detached service, etc.

[8] This would change with wartime losses and deployments.

[9] Four officers, including the captain and 80 other ranks.

[10] V180-V191 were sister ships of which three were sunk, seven surrendered postwar and two survived the Second World War still in the German Navy.

rather close knit as occurs in small warships particularly after three years of close operation. The degree of affection and respect from senior to junior and vice-versa is shown throughout the text.

While a slim work it does provide a valuable insight into the German light forces sailor experience in the Great War. The somberness and matter of fact approach to recounting events rings true.

I have translated the work using the same approach which remains true in conveying Callisen's tenor.

In addition to the contents of the original book, text and illustrations, there is also Callisen's Report on the Sinking of V188, extracts from the German Official History relating the movements of V188 and assorted clarifying charts, Callisen's Career Summary with dates of appointments and rank. For the view from the "Other Side", HM Submarine E-16 Patrol commander's report for the mission which led to the torpedoing of V188. Other clarifying sources are included as footnotes.

C. Corraliza

Contents

Foreword

"Are you on a torpedo?"

Always the same question at home, always the torpedo is confused with the torpedo boat. And every time we members of the "oily art" are shocked to the heart to find so little understanding. Or is it due to indifference alone? I want to erase this – except for the secrecy and seclusion as to the nature of our actions. Everyone knows the artillery, because everyone has seen or heard their shooting, because it has existed for centuries, because we live with it from "Lazy Grete"[11] to "Big Bertha"[12].

We, who exercise on the high seas, hardly meet a strange face, but an old sailor's superstition indicates in our special work that a torpedo itself cannot in a sense be like a dachshund watching a "Bather"[13]. Therefore, be that as it may, it is strictly frowned upon to bring a spectator to a torpedo shoot, as the torpedo would certainly go astray into the ocean bottom ground and thus have missed its target.

It was not until the war that some enlightenment was created after an unfortunate journalist's lovely little

---

[11] A siege cannon so named dating from 1414 credited as the first documented use of artillery.

[12] A heavy siege howitzer used in the First World War.

[13] Reference is to hunting dachshunds that follow their prey without prompting. i.e. Self guiding.

torpedo boat war patrol adopted from the English the completely unfamiliar word "torpedo" and we specialists are served up daily today in every page and leaflet.

The torpedo boat is therefore not synonymous with the torpedo, but only the name of the launching vessel which in the early and developmental years carried torpedoes as the main weapon. Today it is, as we all know, the chief submarine weapon, and all warships from the battleships to auxiliary cruisers carry torpedoes in constant rivalry with Sister Barbara.[14] Thus SMS "Meteor"[15] and "Greif"[16] have destroyed their opponents by torpedo.

The torpedo has gone through the most incredible development stages until finally having, today among all navies in principle, the same appearance and the same means of propulsion. For the first time appearing in the American War of Independence as a propelled Torpedo[17] – approximating a modern floating mine - then continuing as a missile torpedo and spar torpedo to today's self propelled projectile.

There is also a difficult question that no one has properly solved: What is a torpedo? Today it refers to an underwater projectile with a significant explosive charge that moves by its own engine power. However

---

[14] Patron Saint of Artillerist, i.e. torpedoes and ship borne artillery are rivals.

[15] SMS Meteor an auxiliary cruiser, sank H.M. Armed Board Vessel The Ramsey on August 8, 1915 by gunfire and torpedo.

[16] SMS Greif an auxiliary cruiser, sank the Armed Merchant Cruiser HMS Alcantara on February 29, 1916.

[17] David Bushnell's celebrated Turtle used operationally in 1776.

in the early years it was lacking its own power and was merely an explosive device which was dispatched by water currents or carried by a special carrier to the object of the attack.

It was only in the [18]70s when the first self-propelled torpedoes were completed, which by and large were the same forms like today: a powerful cigar shape several meters long, earlier ones from pure bronze today from other, stronger alloys. Basically, it consists of three parts: at front the head with about 100 up to 150 kg explosive charge, then a large air tank filled with compressed air, fuel for the engine, and finally behind the machinery with its countless equipment and valves for controlling the air supply, lubrication, depth and degrees of drive control. The end then forms the propellers and depth and straight running rudder, as we see in the great airships.

This projectile is transported on the ship in long "discharge tubes", carried sometimes above water, sometimes submerged, and "expelled" by firing air or gas pressure, which launches it into the open water. Here it automatically sets all of its machinery in operation and is left to itself. If it does not work the way it should, then the powerless shooter can only hurl a crude sailor's oath after it.

The strength of the torpedo strike is physical damage corresponding to its large explosive effect, which is increased due to the fact that the detonation takes place under water, and in low visibility, evasion is very difficult for the attacked ship. However, the torpedo is not completely invisible, such as is the anchored mine. A torpedo's motor exhaust gases rise to the surface and indicate the path of the torpedo through a clear bubble wake. The calmer the sea, the better it can be seen, and the more likely the possibility for attentive sailors to avoid the torpedo. Escape is no longer possible if the wake is detected too late due to the torpedo's speed of 30 to 40 nautical miles per hour.

This is of course only applicable in daylight, at night the wake is never seen in time to allow an escape.

Presently the Torpedo "boat" acquired its name from only the weapon. Of course it is quite true, the designation boat is no longer valid, and already they have become imposing ships of more than 100 meters in length. With us the nomenclature has not changed from traditional attachment to the early models; our opponents, however, have kept the name for the small 30-50 meter long vessels and call the newer, larger types "Destroyer"[18]. But basically everything is the same. All have a strong Torpedo armament and relatively weak supporting artillery.

The torpedo boat should close to the target as possible and as circumstances permit, to facilitate torpedo targeting and striking the target. Its principal asset is therefore high speed in addition to adapting its outer color to lighting conditions to keep hidden from the eye of enemy scouts as long as possible.[19] Accordingly,

[18] Destroyer is short for Torpedo Boat Destroyer. Essentially large torpedo boats built with a heavier artillery armament to defeat torpedo boat attacks while still having a strong torpedo armament for offensive action.

[19] German torpedo craft where painted in black in the belief it would aid

our element is the night; then we reap as our old torpedo Admiral is so fond of saying, as "the Devil from Hell". When we only wanted to give our opponents the opportunity to do so! Unfortunately, he knows only too well what it yields. He has therefore never been in reachable range at night to look for, and that they have Torpedo-boat leaders[20] gives us an even bitter disappointment. The night attack which we prepared year after year by strenuous maneuvers, meant for us the highest attainment our profession could offer us. But it seemed rather beyond the point if we compare against the duties of war ahead.

---

concealment at night.

[20] Torpedo-boat leaders are destroyers in British parlance.

Beginning of the War

Flotilla Commander: "V188 to go to sea tonight at 8 o'clock"[21]

In front of me is the fellow. It is Sunday, 26 July 1914 at 6 o'clock in the afternoon.

At other times I would have been less pleased about this disturbance of a beautiful Sunday rest, but today I could not quickly enough come to Wik[22] where, ready for combat, the flotilla lay also in peace in the torpedo boat harbor resting after riding out heavy nighttime gales. On my arrival I found the mole black with comrades from the large barracks. It had spread quickly that boats with a special mission should proceed to sea that evening. There, the commanders of the other two boats had already gathered, who received the same instruction as I had.

Brighter future plans are hardly ever forged, higher castles in the air hardly built as at that evening hour. What could be more naturally comparable than our posting with French warships in the Baltic Sea on the occasion of the President visit at the Tsar's court

[21] The 1st Torpedoboat Flotilla with V187 as the Flotilla Leader consisting of two half flotillas. One is the 1st Half Flotilla V191 (leader), V188, V189, V190, G197. The other is the 2nd Half Flotilla G196 (leader) G192, G193, G194, G195. At this time Lt-Commander Willi Wallis was the Flotilla Commander on board V187.

[22] Wik is a district of Kiel and home to the Geman naval base there.

Zarenhof in St. Petersburg and at the royal court in Stockholm![23]

We eagerly awaited our flotilla commander who desired to come along himself. However, this first sortie was to disappointment us: We were given strict orders to not make use of our weapons under any circumstances at meeting foreign warships. But our joy on our first assignment was not diminished: proud heads held high, with the burning feeling of being envied by of all those left behind, we started off to watch ship traffic through the Fehmarn Belt.[24] With downright exemplary carefulness the boat was dimmed so that not even a thin beam of light could sneak into the open ahead and prematurely bring us unintended attention. No spark, no smoke rose from the funnels; like ghosts ships we crept through the night. We encountered another flotilla not paying attention, which recalled from Kiel, met us at high speed - had me evaluating them as if (they were) an enemy!

Already countless nights I had stood my then seven years torpedo boat service on the bridge, far more exciting to follow yet in the war.

The whole crew provided those whose duties were not below deck, none wanted to be pried down from the upper deck. Everyone wanted to see any foreign warships first. Finally, I had to put my foot down and

[23] French President Raymond Poincaré visited Russia in July 1914 using the battleships Jean Bart and France.

[24] Fehmarn Belt is a strait connecting the Bay of Kiel and the Bay of Mecklenburg.

send the watch below deck. Their waiting would have been in vain. The sun rose without a warship to be seen.

As the day was clear and accordingly the Fehmarn Belt could be easily observed by a boat, off proceeded the other two boats; I was left behind. Further, while we lay lonely in the passage and dreaming away a fine summer day, a F. T.[25] Signalman brought me a clear French radio message from the Eiffel Tower[26], sent by the battleship "France" in which the president announced his intended arrival in Dunkirk on the 29th, a special train ordered and more of the same.

Therefore at length the French would not be long in coming; they wanted to arrive on time. And lo and behold, in the morning appeared by 1130 o'clock to the southeast, far below the horizon, the mastheads of two ships. With superior speed they came, two battleships "France" and "Jean Bart"; the small cruiser and the two destroyers were certainly already through the Sound. The "France" leading with the President's flag at the mainmast. A few hours ago came SMY "Hohenzollern"[27] with our Most High Warlord[28] on board from Norway by the "Great Belt". A stroke of good fortune had prevented this strange encounter.

---

[25] F.T.= Funken Telegraphy=wireless Telegraphist

[26] The Eiffel Tower in Paris was used to communicate with French warships via wireless.

[27] Imperial Yacht of Kaiser Wilhelm II

[28] Kaiser Wilhelm II

The next few days gave rise to more peaceful return plans for us. On July 29th I went with two other boats to Warnemünde, where a seaplane week was to be held. We should provide security for the flight tasks.[29]

In a most hilarious and determined merrymaking mood we left Kiel. As the Guardsmen facing the harbor inquired where we were going, one of us answered harmlessly: "To Warnemunde to dance."

How entirely wrong he was, but simply out of this dance, very soon came an armed dance. Already on the 30[30] I received at 6 o'clock in the evening telegraphic orders to return immediately to Kiel. My crews were recently away on leave. So I sent runners throughout the city around which sounded off the call of my boat on a whistle, the prearranged signal for an emergency boarding. The result was that, of course, half of Warnemünde came to the boat and looked at the preparations.

After emergency steam was raised, we ran out and passed the long corridor of the harbor entrance, the moles were solid black with people. Had the bathers guessed the infinite joy their warm goodbye gave to me and my crew, they would have waved and called even more enthusiastic cheers. Often I have since seen how the army transports envies the goodbye that was offered to them on all routes and stops. We should also remain faithful to our profession when moving out into the field: deep secrecy was everywhere. No

[29] SVK – Seeflugzeug-Versuchskommando or Seaplane Testing Unit was based at Warnemünde on the Baltic Coast.
[30] July 30th 1914.

one even noticed anything about our departure later from Kiel.

After we returned late at night from our amusing short trip we found Kiel torpedo boat harbor in complete activity. Equipment still missing here and there was taken on board, fuel issued; everyone wanted to improve his boat who still could.

The next night we went out to Brunsbuttel, where a few days later, we witnessed the mobilization, which set in motion the whole of Germany. Then church bells of the neighboring villages rang a farewell song to us, to many a premature death. We, however, had to remain idle in the struggle until England openly sided with our enemies.

On Outpost

"So foul probably not one of us had thought of this torpedo boat war." "Wait and see! After one week, we are no longer here together. Then we have "loosed" the first decisive shot."[31]

Of course, the foreseeable future absorbed our complete thoughts, as when on a sunny Sunday morning with two flotilla commanders on the dike at Brunsbüttelkoog. Under the pilot house, the old weather-beaten sea dogs stood with their ancient Kiekern[32], inherited from their forefathers, and looked, cursing every leaving English steamer who went downstream. But it was also annoying to see an Englishman run after another. Going against us it is after all. Why not detain their steamers just as securely?!

The fits of anger were carried out for nothing.

The Brunsbüttel passage[33] was far from escapable. And so we celebrated because of the joyful reunion with all these steamers as they headed back upstream to Hamburg.

---

[31] Footnote in the Original "When Torpedo shooting the command to fire: Loose!"

[32] Binoculars. Perhaps Telescopes due to the ancient lineage reference.

[33] Brunsbuttel is at the mouth of the Elbe, downriver from Hamburg where English shipping was attempting to flee from to the open sea.

Finally, on August 4, England also submitted its comments on the matter.[34]

"Dispatch all flotillas!"

Hurrah! Here we go! Tonight we will show what we have learned!

But we had our bill placed without the guest present, because we were far too sure about waiting for him. He would disappoint us badly, very badly.

"Moored on Altenbruch Roads.[35]"

So still not further out.

Well, at the onset of darkness it would surely come!

Just meanness! Full moon today! We could not use the black fellows now.[36]

The command to depart did not occur. Instead we were to anchor along the fort in the channel, the difference was not very much there.

Someone knocks excitedly on my cabin door.

"What's happening?"

"Herr[37] Lieutenant, Aircraft in the Moonlight."

I had not expected that, however, as the first war message.

"Where?"

---

[34] Great Britain declares war on Germany.

[35] At the mouth of the Elbe, between Brunsbuttel and Cuxhaven.

[36] Black Fellows or Dirty Workers, the colloquial expression for Torpedoboats and their crews.

[37] Herr is maintained from the original as there is no proper English term of addressing in the indicated tense.

"Upstream."

"Oh, nonsense. We will probably not hit it."

And so I lay down on the other side.

A flyer on the Elbe River and moreover at night really was an impossibility. -

Instead of our long awaited night attack now followed boring outpost duty.

"The day has 24 hours, and if that is not enough, we will take the night to help out!"

Often we had to learn this eloquent statement in peace, but in war it was our guiding principle.

We were 24 hours on outpost without returning to port, then one day Helgoland, then out again, in that foreseeable sequence. It was no small thing for personnel. Always half the crew is on duty. Below all the boilers are in operation, to be ready at all times for the highest speed, guns and torpedo tubes loaded, and only one officer of the watch on duty.

It's a strange thing when one cruises around isolated and out on a limb, 17 nautical miles (1 nautical mile is equal to 1852 meters) from a neighbor and about 100 nautical miles from the nearest fortress artillery! As well might come to many his own thoughts. There is a tingly feeling - like soda water bubbles, as a lieutenant would say - when you stand guard entirely alone on a powder keg! Powder keg! Yes, God knows when and where the enemy has laid his mines. A fleeting touch and it is over for you! As one has only the small boat, a "dinghy," its 6 men capacity is of little value, and life jackets only prolong the agony.

However surprisingly, one quickly gets used to it and carries on as if it is an unavoidable coincidence.

Nevertheless one cannot see the eggs around underwater - why then concern themselves!

More important is looking out for enemy ships. For this we are here! To make an advancing enemy report in time to allow the fleet commander to take measures![38]

Really, what difference does it make if one or a few boats here go under! Is it any different with army cavalry patrols?

And anyone can harm us at night; no one believes those who wear red piping.[39] After all we are calm. Our peacetime training does not make us as claimed. At night it simply depends on seeing the other first.

And be careful, we were already!

---

[38] The instructions for German orders for torpedo boats forming the securing patrols in the Heligoland Bight read as follows: "DAY DEFENCE."
Object:-The safeguarding of the German Bight in the first place against hostile submarines; in the second place against minelayers.
Conduct by Day:--Torpedo-boats: Not to oppose torpedo-boat destroyers, but give way before them and endeavour to draw them and any other light forces across the submarine patrol lines.
"NIGHT DEFENCE."
Object:-The safeguarding of the German Bight, more especially against mining and blocking operations.
Conduct by Night :-Torpedo-boats : To retire before torpedo-boat destroyers and light cruisers, but resume their station as soon as possible; against minelayers every action is to be taken

[39] Red piping designates Staff Officers.

"One Suffocates Below," groaned coming on the bridge the H.I.[40], alongside the chosen official mess officer (food and beverages that are to be purchased on warships by an officer), war correspondent and skat[41] funds steward.

"It is good for you, for once look at the moon after midnight."

I believe that the air for normal humans originates from below.

The Mess is full of personnel, it is the off duty watch, the gun crew is on the stern. Of interest is the quaint clutter of grenades, pistols and sharp cutlasses located in the officers' areas[42] onboard. Maybe they will need it in a very short time at their battle stations.

"Will you grant me 24 hours company up top?"

"I merely direct you to Herr Lieutenant's[43] attention. I cannot answer for anything else. "

"Well! Making excuses are drawn out long like a straw!"

Also this night is over, and indeed quickly, since we have August to write about.

With dawn comes for us the critical hour: it is too light to attack, you would be shot down before approaching to shoot, and the "trapped" does not sight the enemy

---

[40] Halbflottilleningenieur. Chief Engineer of the Torpedoboat Half-Flotilla. In this case the 1st Half-Flotilla

[41] A popular card game in Germany

[42] On the deck. Original footnote in German text.

[43] Kapitänleutnant in the original

in enough time. Every minute makes visibility more effective, and the opponent on his big ships, if not too heavily occupied, still can send by quite a few large-caliber salvos.

And then it's broad daylight.

The horizon is clear, all around free, no ship to see.

"I pull back my "Salon"[44], I speak to the officer of the watch, I sit down in a corner against the signal pistol ammo box and take under God's open sky "an eye full,"[45] to calm the whole crew.

The "Old Man"[46] shows nothing can be wrong.

A long gentle nudge "purrt" me. My sober orderly.

"When should I bring Herr Lieutenant breakfast?"

"Seaman, no need to wake me up!"

"I thought" ...

"Thinking is a matter of luck! Bring it all the same! "

With a few coated slices he comes up again. This is for me to enjoy after the long night.

"Ugh, How come the bread tastes so?!"

Right, because he forgot to cut off the crust, and the bread still lies in oil, tar and soot from the men onboard, everything sticks to a genuine black sailors[47] fist on a boat.

---

[44] Captain's Chair onboard a ship. Usually used only by the Captain.

[45] Sleep, feigned or actual.

[46] Old Man is the age old informal reference to the commander of a ship.

[47] Black sailors are an allusion to the as referenced perpetual soot which covers torpedo boat crews when the craft are underway.

Yes, should therefore mine be much better?

I would be lying if I wanted to say that. And from the mirror laughing towards me a coal-black face, covered with a mixture of soot and oil from the front funnel, which is only slightly taller than the bridge.

"As far as I am concerned our relief should have arrived now."

"Why do we not maintain [course] a bit for home?" Asked my sole support, the watch officer.

"Searchlight dead ahead," calls a signalman.

"Thank God, there he is! Helmsman course for home!"

His wrinkled face brightens up. ESE 7/8 E [48] is heard. For a long time he has discontinued using that course! Just a shame that he cannot establish it more precisely!

Homeward it is, to Helgoland, so that we first fill the empty stomach of our horse with black coals and then - sleep.

But again a diminishment to our sleep.

“All Flotillas to put out after coaling”

"The radio operator approaches!"

"Telegraphy signalman reporting for duty. "

"Did an important radio message arrive?"

"No, Herr Lieutenant."

---

[48] Footnote in the Original i.e. East-southeast 7/8. A compass has 32 full marks, all read from eighths.

"Ask on the neighboring boat. You might have missed it. "

"Station has always been busy."

"So again an exercise drill," I mumble in my beard - but no – I do not.

After all we drilled enough in peacetime!

Quite inexplicably our flotilla commander never asks us our opinion.

'1. Half Flotilla to the 2nd Squadron ".

"Leaders boat has signaled a communication!"

"I take command! Boatswain's Mate of the watch, clear the front deck! "

The ships sail in line ahead with about 200 meters distance between them. We are on the other side of the line.

Rudder command: "Starboard 20! Both engines ahead full! "

Cursed! I have just arrived there through the gap in front of the pivot point where every ship makes a right, so that the whole line after has another 90 degrees course.

Damn the gap is narrow!

It does no good, I have to turn. If I take another gap, I drive across the bow of the neighboring boat.

The mainmast of the ship behind which I will pass is directly ahead.

"Well done! Follow the compass! "

"Again both engines full speed ahead!"

I sharply look forward and take my guess as it were as I sight a fixed course on my boat.

Still the ship continues turning.

Now it is.

"Starboard 5!"

I turn my boat a little towards the stern of the ship ahead. Now it will be enough.

"Well done!"

With reckless speed we chase a battleship.

100 meters are still between us.

50 meters!

Slowly it turns laterally.

20 meters!

Will it get clear?

Our stem intersects his wake one meter behind his stern.

On the battleship quarterdeck are a few officers and salute - no - today they swing their caps in acknowledgement.

A "Bather"[49] snaps photos.

---

[49] Friendly derogatory term for a guest aboard without maritime experience

"That was for me the most beautiful moment in my Naval experience," tells me the doctor of another half-flotilla, who is temporarily embarked with me.

"Yes, a decent breakthrough is and remains the most beautiful weapon in a torpedo boat."[50]

[50] In a breakthrough maneuver the torpedo boats drive through at 90 degrees to the gaps formed by a line of warships. This was a dangerous maneuver as exemplified on March 4, 1913 when the cruiser Yorck rammed S-178 cutting the latter in two. Other similar breakthrough collisions had occurred in 1905 and 1912.

The First Enemy[51]

Once again we had our five-day guard cruise behind us and lay stopped at the Jade. Mirror-like sea. Wonderful weather. Once again, day dreaming on the bridge and far away from thoughts of the dangers of war...

But the reality ... Even the entire bridge staff is startled by the whistle blast on of radio shack that takes us back to the world of facts.

"Enemy submarine southwest Helgoland." "The Devil may know, I feel as if we had not yet currently come to W'haven[52], dead certain we would be engaged to hunt."

Just as we intend to turn into the Jade, here it comes! Instead of left we make a right and head out hunting. Although we had often hunted, but had never seen one. Many a submarine commander watching us through his periscope and enjoying our blindness.

What does it matter! What is commanded is done.

"To the guns!" A voice command and not using the alarm bell. This is only for urgent cases as when everyone available is to leave what he has in hand and hurry as fast as possible to his battle station.

---

[51] The events in this chapter correspond to August 20 based on the War Diary or KTB(Kriegstagbuch) in German, of the 1st Torpedoboat Flotilla.

[52] Wilhelmshaven. Abbreviated in the original.

"You know the announcement! Ten Marks[53], whoever sights him first! "

Revived pairs of eyes stare into the water, sometimes far and sometimes close.

There! All snap their heads around. My section leader opens a wild machine gun fire.

"Signalman! Semaphore Message: K. to K.[54] What do you have."

Answer: "K. to K. I shot at tin cans. "

"K. to K. You are ..."

But he was not entirely wrong. The bad habit of throwing the empty cans overboard has led to some alarm and many evasive maneuvers. At large distances such a box is much like a periscope.

Enough, our hunt went on, and everyone swore at the unlucky man, who had only seen this tin can, to whom we owed our trip because of his report.

Suddenly my look-out indicates in the water on the starboard side and shouts: "A periscope".

What cheek of the fellow!

Slowly it rose out of the water to a height of one meter, not 15 meters from the boat.

Boom! My first shot of the war had come to pass. Awful Shot! We were too close beside it; the shot went

---

[53] A Mark is a German unit of currency.

[54] Footnote in the Original *Commander to Commander, Kommandant in German*

just over it despite the cannon muzzle at maximum depression.

The submarine commander would have made a surprised face when his observation port was sprinkled by the splashing waters of the shell. He turned his periscope around immediately, and probably saw in front of him the high hull of my forecastle before pulling the periscope down amazingly fast.[55]

By rudder and engine maneuvers I presently tried to ram the submarine. But without success.

How should I! Our boats run submerged at about three meters, however even a submarine with an extended periscope at the top of his tower is at least four meters below the water surface. We can neutralize a submarine only if it has a fault in the diving equipment.[56]

At least - we were the first of the flotilla who had actually seen the enemy, and fired a shot at him.

And I have never more joyfully paid ten marks.

[55] Either H.M. Submarine E 4 or E9, both reported being hunted by German torpedo boats on August 20 while in the vicinity, .i.e. SW of Heligoland.

[56] Callisen refers to being unable to dive as the desired diving equipment fault. Germany only began using depth charges in 1915 as a means to destroy submerged submarines. Ironically, no British submarines were destroyed by German depth charges in WW1.

Baptism of Fire

August 28th 1914[57]

There was little appreciation on the bridge for the magnificent sunrise after twelve hours lookout duty. Probably every five minutes we passed the glasses to our eyes to scan the surrounding nocturnal horizon for smoke or the dark shadow of hostile invaders in the German Bight.

Yet another twelve hours, then the outpost duty was over once again, and then one could sleep again.[58]

For now one had to be satisfied that the weather gods meant well with us. A slight swell rocked the boat gently up and down quite suitable to lull oneself. Of course, the blood-red sunrise promised no nice day[59]: Rain or hazy weather.

Ulli[60], my watch officer, my faithful support through three years of peace, had morning watch. A thought to exploit morning coffee does not mean, as usual, to

---

[57] The ensuing engagement is known as the Battle of Heligoland Bight.
[58] The 1st Torpedoboat Flotilla had taken station at 0430hours, arrayed in an outward facing curve approximately 35miles from Heligoland forming the outer patrol ring defending the German Bight from British surface ship intrusions. The boats were arranged in order from S/SW to N/NW as V189, G197, V191, V190, V188, V187, G194, G196, and G193. Naval Staff Monograph. Battle of Heligoland Bight. p. 122. See Appendix B in this book.
[59] Seamen saying "Red sky at night, sailor's delight. Red sky at morning, sailor take warning."
[60] Senior Lieutenant Friederich Ulrich (class of 1906) reported to V188 in 1912.

drink it while standing on the bridge, but relax a moment in the cabin, to gain new strength for the day.

But Mars had determined differently about my breakfast.

I have not yet sat still when I noticed from the boat vibrations that the engines had gone to full speed.

No sooner had I arrived on deck, there came before me the runner with the message: "Submerged submarine situated right ahead!"

Right, because it was an estimated 6,000 meters away.

"To the guns!"

For the first time the alarm bell rang in all areas of the boat! How they came out, the recently still bleary-eyed figures. All tiredness has vanished. Even my “Bather”, the H.I.[61], came out with unusual speed, to see the enemy.

And all those whose battle post was below deck in the machinery, the boiler room, and the magazines. What emotions may they have controlled? How we are to be envied, we stand up and see what's going on! What tremendous, blind faith in the commander must there be, that everyone down there keeps calm and is confident in victory! I have often thought about it, and it has always closely connected me with my crew.

An order to the radio shack. Radio message to the flotilla: "Enemy submarine in sight" The time passes a

[61] The Half-Flotilla Engineer.

while since any whistle from below, and instead of the message: "radio message delivered," sounds - I hardly trust my ears - "radio message cannot be delivered, the transmitter has just blown." So, the first warlike radio message was miserably ruined.

4000 meters.

"Permission to Fire!" No sooner stated yhen already whistled over an iron greeting to the submarine.

"Short, middle!" A powerful spray of water and black powder smoke of the bursting shell concealed the target.

But if only a gust of wind would dissipate the explosive cloud!

Finally, the target was again visible, but only the upper part of the tower still stood out of the water. The boat was diving.

Now we had to focus on the diving site with binoculars and watch the periscope constantly, so we did not get in front of his tube and he answer our cannon salute with a torpedo.[62]

We approached him at high speed hoping to neutralize him by ramming. But we were hardly 100 meters away, when he pulled down his periscope and disappeared, never to return.

While I was drawing circles around him and did not give up hope, the radio shack whistled.

[62] Some types of British submarines had fixed beam torpedo tubes fitted in addition to the traditional fore and aft tubes.

Message from northern wing: "Enemy cruisers have broken through!"[63]

At last they were here, that which we had been waiting for four weeks! And only us outposts!

The weather was still hazy. One could see barely 6000 meters. Of my neighbors who were about six nautical miles off, we had no idea.

In daylight as a single boat without any armor protection attacking the enemy, would be like competing at a tournament using a needle as a weapon against a knight in armed combat. Therefore course for Helgoland! On the way one of the neighbors will soon come in sight.

"Helmsman report when we are half way back. Then, we want to make left turn and look at the enemy ourselves. "

Inasmuch as reported Ulli "port beam two British cruisers!"[64]

My answer was not very parliamentarian, because now our way to Helgoland was barred. And what is unpleasant about the matter was due to the submarine excitement my boiler fires were not cleaned and therefore the cruiser steams as fast as we do.

Searchlights by the British, in any case, ask whether we are friend or foe.

---

[63] Sighting report from G194 received about 0730 hours

[64] HMS Nottingham and Lowestoft. Time is about 0750hours

"Shall we not reply?" Asked Ulli.

"Yes, now we want to "watch suspiciously" make LL" (The peacetime signal at the start of target practice.)

It was successful. The English had the searchlight signal for the correct answer and takes me for one of them. Their shells were waiting in vain. Otherwise it would also undoubtedly go badly for me. After all, I was near her, about 5000 meters, still quite uncomfortable.

We better go further south.

Then come two more out of the fog, bearing in front of the other two. The searchlight game begins anew. But this time without success for me.

As soon as we have given our response, as it flashes over, we come under fire from the vessel leading the other three.

Seven nautical miles from land. Well, running away probably seems futile. Because is it any better: Is it!? Undoubtedly you want to, if fortunate enough, to take one with you with a torpedo hit, rather than merely playing target and observe the flash of volleys and wait for when it hits. Uncomfortably cracked the English shells when exploding, and the hideous green appearance of their explosive bursts.

I am searching again the horizon, since none of us comes into view. I expected my left neighbor.[65] Innocent, he unknowingly continues his outpost duty.

---

[65] Left hand neighbor boat is V190. Lieutenant Wautiz commanding.

Because of my currently damaged transmitter I could only receive, but not send radio messages. So I Morse over with the searchlight: "I am being chased by enemy cruisers."

He is presently on my course, but proceeds very slowly. He must not have understood my searchlight message, he must indeed be blind, that he does not see the shell splashes around me.

I hurry quite close past him, calling out to him my observations. A shrug is the answer. He cannot do more; one of his boilers is failing.

The Englishmen, which in the meantime had been joined by nos. 5 and 6, although only for five minutes, directed their fire on my crippled neighbor, probably in the belief that the prize to be caught was closer to them. They shot very well on him.[66]

Impacts three meters away from him - now he will remain behind! - But no, his boilers were in order, and he is steaming even faster than we do.

When he came up close, the shells salvos poured again on both of us. One shell landed 20 meters from my stern, a splinter about forearm length flew in tangled twists and turns over us and slapped across the bow into the water.

This wild chase lasted 45 minutes, without having either of us receiving a hit. Then the Englishmen

[66] The Royal Navy's 1st Light Cruiser Squadron comprising HMS Southampton, Birmingham, Nottingham, Lowestoft, Falmouth, Liverpool

turned off and disappeared. Why, to me, is still inexplicable.[67]

We had peace, but it should not last long.

"Off Port bow ahead, three destroyers, course southwest." Thus crossing past in front of us.

On the starboard side the other three boats my half-flotilla had now come into view,[68] who would help us if it became precarious?

---

[67] An eyewitness account from onboard HMS Southampton records the event as follows "We acted on the good old maxim of going where you hear a gun, and stood over towards the firing. It was impossible to see anything, but at the same time it was undeniably a most thrilling sensation to be moving through the mist at 24 knots towards the first sounds of gun-fire in battle that most of us had ever heard.

At 8.25 a.m. two black shapes, which revealed themselves to be German destroyers travelling at a very high rate of speed, appeared on our starboard bow.

They had evidently been patrolling seawards, and, hearing the firing in between themselves and the German coast, they were scooting into their homes as quickly as they could.

We got the forecastle and starboard-bow gun to bear on them and opened fire, but, as the mist prevented any ranging, we could only hope for a lucky hit.

Two white puffs or splashes were seen to proceed from the enemy, and it was not until some ten minutes later, when three witnesses saw the track of a torpedo across our stern, that we realized that the Germans had fired two torpedoes at us.

The hostile destroyers were going at least 32 knots and were moving between enormous bow waves, with their sterns tucked well down, and in about three minutes they had crossed our bows and disappeared in the mist. " from p. 51, A North Sea Diary 1914-1918. Stephen King-Hall. Newness: London 1936

[68] V189, V191, G197 at about 0825hours from p. 123 Naval Staff Monographs III,Ibid.

Relevant extract from the German Official History is provided as an Appendix.

The British turned to a reciprocal course, to engage in battle.

"Permission to Fire!"

Our greeting was returned most vigorously. As I drove forward, The English seemed to think I'm the leader, the one to deal with first.

Psss, Psss - so the shells passed continuously over in front of the bridge.

But this was a different case, because now one could shoot back. And we also had the grim pleasure to watch two hits on our part.

Also I myself received nothing this time, though the enemy had expended an estimated 150 shells on me.

But the major achievement I took home from these skirmishes was the conviction that my crew had magnificently passed its baptism of fire. Henceforth, with such a crew I could do anything.

But this day subsequently brought me a great sorrow. Ulli[69] was assigned elsewhere to provide for an unusual and more important position.

I received a new officer, who had just returned from South America - to find an early death. But more of that later.

---

[69] Lt. Ulrich later commanded the submarine U101 in 1918 and survived the war.

Storm Surge[70]

Is there anything better for a torpedo-boat commander than to berth with a wind force 10 on the windward size in Helgoland harbor?[71]

Is a little self-confidence not justified when one has mastered his black steed against storm and current? As easy as it appears, it in fact does not apply to a torpedo boat, where the front and rear has only two meters of space.[72]

I think that only a torpedo boat driver can assess this properly. It looks so simple and yet is only learned through many exercises, and even then only with appropriate natural abilities.

What do you large ship sailors know of maneuvering? What do you all know about seafaring navigation, of oil skins and "Cupids", of pea soup[73] and "torpedo boat pepper?"[74]

---

[70] This chapter corresponds to entries in the 1st Torpedoboat Flotilla War Diary dated September 18, 1914.

[71] A Force 10 is properly a storm with wind gusting between 55-63mph and waves 29-41 feet in height.

[72] Callisen refers to the berth width being two meters width.

[73] Pea soup is the vernacular for thick fog.

[74] Footnote in Original, "cupids" are tall sea-boots, and "torpedo boat pepper" is called the soot from the chimneys. Since the galley is forward, the food must be transported on deck in any weather. It is very easy for soot to enter the bowls.

High as a tower rose the splashes when the breakers hit the breakwater on the Southwest mole. The white spray was widely scattered across the inner harbor.

Higher and higher rose the sea, as if to certainly shift the massive concrete blocks and rail and tear us from the snug shelter we found.

At three we lay side by side behind the mole and each had applied our steel anchor chains as mooring lines.

The high tide period was over. With each falling centimeter the boats appeared less high above the mole and progressively reduced their surfaces exposed to the storm.

Night fell in the period of low tide. So the whole crew was sent to rest less duty postings without exception.

6am would be high tide again.

But in the meantime, at least the storm had let up and turned to the North West, allowing us to lie undisturbed behind the island.

I ordered to be awakened at 4 o'clock. The weather gods had played a trick on us.

So again, the new port installation was to endure flooding and discard proof that they could resist all attacks. But this time it should be even worse - and even more impressive to see because water had been building up in the German Bight.

I was the outer berth boat, so therefore must cast off first if the lines should break this time. So I had to make ready the engines and remain on the bridge.

The tide rose, and with it the exposed boat surfaces to the increasing wind.

It still had not reached its peak when it was already alive in the port as well.

Loudly snapped the cables applied to our rear neighbors, they ran into the inner harbor, which was actually already crowded due to a flotilla that had arrived before us.

Then followed our lines.

With great difficulty I brought my boat alongside a collier, who took me under his wing. We held using thick as your arm hawsers and strong fenders.

Others ran off and stayed outside to endure heavy seas by themselves.

The recently arrived boats in the inner harbor very soon experienced trouble. They had felt far too safe in there. And that is why they were in Morpheus'[75] arms and delighted with this lovely opportunity for once to thoroughly sleep.

Crash! Here too snapped the cables.

And then an image as can only be provided by a flash of lightning.

I can still see one of the commanders in front of me dressed in nightgown, hat and loafers. And often he has in his peculiar war garment furnished material for our laughter.

---

[75] Ancient Greek god of dreams and sleep.

The view of the harbor from my bridge was incomparably beautiful.

The sudden impact of green walls of water on the moles rose up high as a house in the air, only to then scatter in a white spray of thousands and thousands of fine droplets.

Long will this night be remembered by the naval gunners, who had their posts on the pier, until at long last, the Helgoland lifeboat, which has proven itself so many times, received orders to pick them up.

They probably could not stand there for long. Breaker over breaker fell here over the mole. Even today it cuts me to my heart, when I think of a small sailing yacht broken loose from its anchor and smashed to pieces on the breakwater.

But the Mole held. German technology had once again proved that it can also even defeat the forces of nature.

Helgoland, you red rocky island, probably know your worth now is higher in wartime than we torpedo boatmen. Woe, if you were still British![76] Only too monastic have you become, since everything that was feminine was taken away at the war's outbreak; and now, the only female creature who stayed there and like a knight's wife of old lady observing from high balconies the bustle of the harbor, the young lady lion of Hagenbeck,[77] was carried off by an airplane to Norderney .

---

[76] Helgoland was British owned until 1890.

[77] Tongue-in-check reference to the sole female on display in Helgoland being

Hold. A "lady" nevertheless still remains! Although not made of flesh and blood, but no less popular. The lady at the cinema! And when she appears, receives everyone's Cheers!

How lucky are you over there on the mainland to be envied! But it is a given with you guys - wherever the sun rises – are provided real women!

---

akin to the female lion on display at the Hangenbeck Zoo in Hamburg, i.e. to observe but not touch.

V188

In Distress[78]

"Vessel ahead!"

As always; so many pairs of eyes on the bridge, so many binoculars! A vessel must indeed be there; but how and what? It could not yet be accounted for.

"The Flying Dutchman in incarnate form."

A picture of misery! It had once been a stately sailing ship, now two poles sticking up out of a hull deep in the water, yards in a disorderly mess and in fact the sad remains of a former sailing vessel.

It had to have been suddenly surprised by the storm. Nearly all the rigging - the thick ropes that frame a sail - stood, but the canvas was torn away. Some tattered remnants testified to that past glory.

Closer in, we recognized the Swedish flag at the foremast, but it was entangled in a knot, and that means. "I'm in distress, requesting help" when all other signal flags are carried away by the sea.

Now it steered with a small reserve sail straight towards land in order to at least save the crew of the sinking ship.

A miracle that it had been spared from the countless mines on its haphazard crossing! Into the Skagerrak

[78] The events in this section occur on September 19th, 1914.

he hoped for and now drifted before the wind to Borkum[79].

It was only two miles from land; we arrived just in time to keep the skipper and his "Anna"[80] from Brantevik[81] ahead of the worst. Amongst my officers was an old sea dog, who had gone many years in Nordic waters and provided the translating, and so we learned all the misery.

Not only had the “Anna” lost her sails, but she had also sprung a leak. The crew still held her buoyant with continuous pumping. But how long will it before the swelling bran cargo bursts the hull apart, for then the heaving burden and worries behind them, now remains only the small dinghy for the crew.

Semaphore message from leader boat: "Stay in the vicinity of the sailing ship; I will summon a tugboat."

Our reply, "May I take it in tow?"

"Yes, if you can."

Nevertheless this doubt of my seafaring skills then awoke my whole ambition.

It was certainly an uncomfortable business to have to lie still for a time with the engines stopped, because a better target could not be offered to enemy submarines.

---

[79] An island at the mouth of the Ems River

[80] Anna was built in 1885, 204 net tons, 37.6m x 8.1m x 3.2m LWD. She was on a voyage from Poole to Copenhagen when encountering the gale mentioned in the previous chapter. Schooner rig. Three master. Apparently one dismasted in the gale.

[81] Swedish home port of the Anna.

But still! Waiting would mean to lose my Swede as he drove fast to the shore against the wind and sea.

"Clear to tow!"

The command was sent to the sailors. Hey, what life came into the flaccid members of the crew! All had given up, already taking leave of the old gallant “Anna” and now we should still provide help!?

No better tow maneuvers have succeeded in my long torpedo boat service. But the Swedes also worked feverishly.

It was high time; the water was already shallow, since we increasingly approached the shore.

I slowly steam ahead; now comes the critical moment, whether the lines keep or break the stiff tackle. And it was a heavy, clumsy vessel there behind me!

The line holds!

Using slow speed I tow the "Anna" to sea.

There is a strong swell. Will the line continue to hold?

Here comes in the distance a tugboat, an "Anna", but a German one.

If the tow line breaks now, it does no harm. We are far from the shore so that there is enough time for the tow to take her away.

No doubt we can probably steam at a slightly higher speed.

Crack! After a quarter of an hour, in an especially high seas where the "Anna" is pulled back, the line broke.

The joy was so short, but this span of time was enough: The life of the Swedish "Anna" and perhaps her crew was saved. It seems doubtful whether its complement could have attained dry land in their small dinghy amidst the high surf.

On the Swede there was a universal shoulder shrug at the moment the line broke and back came the same old resignation to their seemingly inevitable fate: We are just not going to help.

I steam alongside, and convey to the captain that a tug was already in sight and would stay with him to save the crew if the ship could not be towed.

To his relief, I remained there until the two Annas had each other in tow. Then I hurried again after my flotilla.

A few days later, there was in the Order of the Day an acknowledgment of praise for the commander of the German "Anna". He had brought his happy namesake to Bremerhaven.

Whitby Hartlepool

Unsuspecting we lay at anchor to anchor and played our four man game of skat, in eternal anticipation of what was to come. Then the door opens up, and walks in a sailor with a letter. In itself, nothing significant.[82]

But on the envelope was:

"Open 2 hours after departure."

This sounded very mysterious and gratifying.

There was something in the air.

"Well, gentlemen, I think we take" an eye full" to stock up. Who knows when we weigh anchor and how long do we have the pleasure of it. "

A torpedo boat man can always sleep, in all circumstances, standing, sitting and "coiled up"[83] in some corner because he is tired all the time.

Undressing of course, was unthinkable for at any moment could come the signal come to weigh anchor.

The night was pitch black.

It was not until I go again on deck and look myself at our position to the leader boat. All cats are grey at night[84], as the small insignias cannot be seen, in itself

[82] The date is December 15,1914

[83] Original footnote, "coiling" means pooling a rope in a circular manner on deck, the expression is transferred to people who huddle together when sitting or lying down.

intelligible only to torpedo boat men. And immediately identifying the leader boat when anchor is being raised is extremely difficult. Then I retire to my cabin and a hard bed.

Bang! Bang! Knocking on the door.

"What's happening?"

"Herr Lieutenant, anchor is raised!"

Oilskin on, cap, gloves in hand, then on up.

As chicks around a mother hen, so the boats gather around their leader boat and try to use the shortest possible route for the correct position, None too easy however if one obstructs other boats.

Where to now? The half-flotilla attaches itself to a cruiser.

Aha, this time there is something special. If only the next two hours were to only come around and one could open the mysterious letter!

Finally, finally!

The navigation compass flies into the envelope and tears at it.

There it is: Against England!

And we clearly at the very forefront!

To Hartlepool, on the Greenwich meridian, wanting to cross over in itself simply had to be termed audacious!

"Guys, we shall chase the English out of bed!"

---

[84] Proverb from John Heywood's Dialogue of Proverbs, 1546.

Within seconds it was over in all the crew spaces, the sailors were grinning and already brushing up their sleeves, as if off to a wrestling match, the stokers spat vigorously in their hands to hold their fires clean and strong. Now it was smoke and sparks as we steam off!

The weather was kind to us.

Smooth as glass the usually harsh wintry North Sea! As the sea voyage was already bearable, one could sit on his lantern box and dangle his legs, and frying eggs did not jump from the frying pan!

There was even hot lunch! And afterwards "Negro Bouillon" with manatee milk. [85] Admittedly, for eight years I share only crew fare with my people. Of course it is especially unimaginable that it is cooked in the small space available for the officers, and there can be in the pantry just a man standing, and also he must set the dishes between his feet when washing up.

Nevertheless, is there in such weather anything more beautiful than the seafaring?! Plus the prospect to finally get at the enemy!

Gorgeous it was, and then followed by a deep black night, such a proper torpedo boat night as we black fellows could not better wish for.

"Herr Lieutenant," said my auxiliary coxswain, a true Mecklenburger, "To the west of Dogger Bank the weather is bad. I do not like the wind at all. The sea begins to be rather heavy. "

---

[85] Footnote in the original. Coffee with condensed milk.

"Do not paint the Devil on the wall."[86]

However, he would be correct in his prophecy.

At 4 o'clock in the morning it was already quite uncomfortable, and the closer we got to the coast, the worse the weather. All northerly winds, so that the coast offered us no protection.

Early at 8 o'clock we were very close among the rocks of Whitby.

The thrill was indescribably beautiful.

Having the same crew on the boat for three years one certainly expected to be superior in combat! Only the weather was a downright beastly. Because of which we torpedo boats could not take part in the shooting. But we still had heartfelt joy to be there.

We clearly saw the flashes of the German salvos. The houses brightly illuminated by gunfire, garishly greeted us, so close were the ships to land.[87]

"Good morning, Mr. Grey![88] You will only receive a fit of anger if our greeting is delivered with your breakfast."

"Britannia rules the waves!"

---

[86] A common German expression. An answer if one is reminded of the worst possible consequences or happenings which may occur.
[87] Approximately 1150 German artillery shells were fired in 1hr 20min commencing at 810am local time.
[88] Reference to Edward Grey, British Secretary of State 1905-1916. Delivered Britain's ultimatum leading to war with Germany. Reputed to have made the memorable comment at the war's commencement, "The lamps are going out all over Europe, we shall not see them lit again in our life-time".

"On this joyful day I will go below for breakfast to celebrate. Report to me immediately if any problem occurs."

In oilskin and sea boots I sat at the breakfast table in the sofa cushion, right next to me where what was left of the butter dish, sausage tips, and bread basket. My fellow whose chief merit was not to become angry in the nastiest weather, held a coffee pot and cup with a practiced knee bend.

Sufficient to say there was nothing on the table; nothing would stay long on it.

I scarcely have the first bite in my mouth, when a runner comes with the message:

"No. 3 has a man overboard."

That's the limit! 15 nautical miles from the coast of England, wind strength. 8

I was up! Struggling to get to the bridge over cluttered piles of coal sacks. As will happen when one is slightly wet and running all out in seaboots!

The no. 3, my front man, experienced the loss but also noticed and immediately turned about.

It was a petty officer, when he came out of the boiler room companionway was taken by the heavy seas and washed overboard.

A lucky child, as there is rarely any other.

He had been wearing little gear in the boiler room so he could stay afloat despite the high seas.

The boat went up to him, very close, and then– an approaching rolling wave took him and threw him back on deck. Strong sailors' fists grab and hold him so the same wave does not take him back into the sea.

And high up in the mast flies the signal: "The man is saved!"

Age well brave petty officer, you've done well!

However this incident brought us back starkly to the current situation, as now the British knew our presence due to the bombardment, to return home as quickly as possible in order to be taken not by superior forces.

At full speed departed the other ships.

"The senior engineer is to come to the bridge."

With difficulty he worked his way up.

"Chief Engineer, how is it with our coal?"

"Herr Lieutenant we can no longer maintain full speed, I only have a few tons in the operational bunkers. The whole deck cargo is lost overboard, and trimming from the reserve bunkers cannot be done in this weather. "

"We have to match the flotilla, and then continue with the voyage at 20 knots. Will it then be enough?"

"Yes, it's enough."

Hardly had we caught up with the leading boats, as a salvo hits before us alongside the half-flotilla.[89]

---

[89] See Appendix C for a map of the movements of British and German forces

Well?! One could see nothing behind us.

The forward boats turn south, we follow.

Damn it: At the front a small British cruiser peppers us briskly.[90]

The Ensign[91] on watch straightens up.

He beams. Now he gets to see something of the war.

The enemy also turns to the south. Attacking him is impossible in such weather. The torpedo tubes are not to swing out. Since one simply can run away using high speed.

The Englishman shoots poorly. The shell splashes are far apart. For in rolling ship even the best gunner can hit nothing.

Suddenly the enemy turns against the seaway, driving north at high speed.

His bow lifts high out of the water as he steams on against the sea.

The reason of his maneuver remains inexplicable. But what does it matter to us! We turn back to a homebound course.[92]

---

described in the following pages.

[90] The Town class cruiser(s) were actually larger and better armed than their German counterparts. The small or kleiner identifier in German is to distinguish from an armored cruiser. The Towns would be termed as light cruisers today, a post-WW1 creation with the introduction of heavy cruisers.

[91] Fähnrich in German naval rank.

[92] An eyewitness on HMS Southampton records the engagement and withdrawal here: "The light-cruisers' screen was spread as usual in groups of light cruisers about 5 miles apart ahead of the battle-cruisers. We were the southern group, and the Lion was about 8 miles on our starboard quarter. But a quick look round as I ran up to the after-control revealed nothing except the

The boat is harshly pounded, jumping from sea to sea. The high speed buries the bow deep in the water only to rear up in the next moment and throw masses of water over the bridge and funnel.

The Ensign looks at me questioning.

Birmingham shrouded in driving mist about 2 miles on our starboard quarter, bearing down to our support, and the two groups of Huns, which consisted of three light cruisers and a dozen destroyers on the starboard bow, and two light cruisers and an armoured cruiser (the Prinz Adalbert)
and destroyers on the port bow.

We went straight on at 25 knots, head into the sea, and spray flying over the ship in sheets.

The Huns came straight on, with the sea behind them and the destroyers bobbing about like corks. As both the enemy and ourselves approached each other on opposite courses, it appeared as if we were about to pass between the two groups of German ships. In fact, both broadsides were in action for a short period, but when the group on the starboard side were bearing about on the bow, they altered course approximately seven points to starboard and stood across our bows towards the wakes of the other party, which by this tune were bearing on our port quarter. To conform to this movement, the Birmingham having got astern of us in support, we altered course to port, and steered parallel to the two groups of enemy. The foremost group slowed down, and the Germans then assumed one long straggling line which extended from our starboard bow to our starboard quarter, the mean range being about 6,000 yards. Fragments fell on board, but they never hit us, which was a poor display for five ships.

We opened fire with all guns bearing, but the gunlayers became confused at the number of targets and each gun was firing more or less independently. This, added to the fact that owing to the sea and spray the telescopes were useless and they had to use open sights, made accurate shooting impossible, and I don't think we hit any Huns, though we managed to straddle the armoured cruiser. We were, however, recalled to the northward, where the Lion was, as any minute the German battle-cruisers were expected to come out of the gap, and as we were short of T.B.D.'s the battlecruisers wanted light cruisers with them in case they were attacked by the German destroyers. Stephen King-Hall Ibid pp 76-77.

Though this goes against the grain for him, this unrestrained exposure of the ship.

"Help nothing, Berger, from this cauldron we only need only a time out."

Crash! Again a green sea pours over the bridge. Splashes smashed the 1cm thick window panes and gives us a proper slap in the face. The helmsman appears studded with shards of glass. I also have a few driven in the cheek.

Pity nothing! We are used to this grief.

Said yet one of my half-flotilla commanders, after he once met a fellow from the Marines with a shard in the eye after such voyage. "This does not impress me with any such sort, my helmsman has at least twenty in the face."

The chief engineer despairingly arrives on the bridge. "During this voyage the coal is not enough to reach home."

"Helmsman, how far is it to Zeebrugge?"[93]

"Exactly as far as to Helgoland."

"Hourly coal report! As soon as the weather permits, trim off the reserve bunkers! "

"I'll work out times, how far we get," says the H.I.

A damned situation.

Message from the radio shack: "Strong enemy W.T. traffic."

[93] German naval base in occupied Belgium.

Naturally! Now they hunt with anything that has screws[94] coming after us and intend take us. But we do not want to make this enjoyable for them.

The H. I. comes up again. "Up to 8 o'clock this evening we can continue, then we have to go on half speed to arrive at Helgoland."

"How is it with the engines?"

"Then at 10 o'clock we need to decrease and make engine revolutions for only half speed."

"Thank you very much!"

8 o'clock.

"Radio shack, do you hear heavy traffic from the enemy?"

"Yes."

So we continued with the same speed. Now I have no choice but at 10 o'clock to separate from the other boats and to build my own lucky star.

"Portside engine stop! Starboard engine half speed ahead! - To the engine room: Carefully go smokeless! Turn off two boiler fires!"

With a heavy heart I let my half-flotilla drive off. But I have always been lucky!

Never have I sailed smokeless like today, never has the crew kept as sharp a lookout.

---

[94] Screws are a colloquialism for propellers.

There is still proven experience: Whoever sees the enemy first at night is the superior. And I could still shoot!

Endless hours creep by. Will I reach Helgoland? Or do I have to anchor at sea defenseless, exposed to every submarine?

Damn the devil! If only we do not have to play English Tennis tomorrow!

Since now there are flashes up ahead!

The light of Helgoland!

God be praised!

Around 2 o'clock I was in port. I stood on the bridge exactly forty-eight hours, and the last fourteen hours had no "cocoa"[95].

I had three tons of coal, a teaspoonful for the large appetite of my boilers![96] With difficulty I reached the next mole. At the coaling mole I could not steam any further.

With leaden limbs I went down and had the duty watch officer use lines to bring us over to the coaling station.

But I slept - slept - slept - twice around the clock...

---

[95] Energizing refreshment presumably coffee or similar caffeinated drink.

[96] V188's coal capacity was 136tons. According to the Official German History of the War at Sea, V188 reported 6 tons of coal and 2 tons of oil remaining upon arrival at Heligoland. The Official History estimated V-class torpedo boat capacity for this operation at 162 tons bunker coal, 20 tons coal as deck cargo and 54 tons of fuel oil. Oil was sprayed onto the coal to intensify output.

The next day all was forgotten. What remained was simply the pride of having been to the western longitude.

Christmas

Christmas was on the calendar, and our festive mood was unchanged, only not ebullient over thoughts of our loved ones at home.

In the coming commanders' meeting we were unanimously in favor of painting the telltale red calendar over the black. But the German Michel[97] sat too deeply in our blood, and we wanted to make a little pleasure for our people.

We were lucky on December 24.

The flotilla was anchored in the estuary, undisturbed by the official and unofficial bustle on land; we were entirely among our "immediate family".

Only our outpost duty was to start by 4 o'clock. So we had until then to be able to celebrate an undisturbed "Christmas Day" and could count on a few hours of rest.

"I bet the British will pay us a visit tonight."

But nobody was against it.

So at 12 o'clock noon I gave a short, well, let's say seamanlike-warlike Christmas sermon, and then the sunset was accentuated.

---

[97] Figurative representation of the German people. Similar to Uncle Sam for the USA and John Bull for the UK.

In broad daylight, you cannot possibly light a Christmas tree! So "dim it", i.e. Fold down the metal shielding against all the windows. Only this should not prevent light exiting outward, as happens at sea at night, but now the inverse: The sunlight must enter whether it was blocked off or not.

Individually I let people come into the cabin to give them a little memento of this wartime Christmas: a little package that I received from relatives and acquaintances. As I stood in front of my little gifts I envied the cruiser next to us with disapproving eyes. They all had their sponsor cities were almost too spoiled, and we who eke out our lives under a number[98], had nothing, absolutely nothing.

But yet I hope to have reasonably met the appropriate requests.

Only one thing I could not substitute for my people: the missing mail from home, most of which were still outstanding.

Then I walked with my officers through the narrow side compartments, all of which had their little Christmas tree, as I imagine the example with the Lilliputians.

"Well, what did your fiancée write you?"

"Nothing, Herr Lieutenant."

"I would end it".

---

98 German cruisers were named after cities-Emden, Hamburg, Bremen, etc and so had a relationship with their named city. Torpedo boats were numbered, as in V-188 and had no patrons as the author refers to regarding larger ships.

There are two- having a look at the "Week"[99], and looking very carefully at a photo.

"Who is that?"

"Hindenburg[100], Herr Lieutenant."

"What does he look like?"

Then the second: "Well, I do not want to be on report."

Finally we four officers gather in the mess, three worthy spouses in their thirties, plus a cadet. To celebrate, I could probably give up my lonely existence as commander and dine with my officers.

The mess committee allocated something very special for us: roast goose. Something unheard of in torpedo boat conditions, both in scope and style.

As always, I'm sitting on the black horsehair sofa, which combined with a curtain otherwise is living, dining and bedrooms of an Ensign.

In the companionway two legs are visible descending, followed by a shirt, soup bowl and head.

Today, it is to be elegant for us! The soup outside brought and offered in plates.

"Man, are you not burning your fingers".

But my staid fellow cannot be upset. Faithful-hearted, he replies, "It is not so hot."

---

[99] A weekly. Newspaper or other periodical is presumed.

[100] Paul von Hindenburg, German Army Field Marshal. Recent victor at the Battle of Tannenburg resulting in much fame and media attention.

A pop outside. For once, today there is "Schuum"[101]. A bottle - more is not allowed; we still have a long night ahead of us. But we have to initiate our new members, with, in the absence of ice, nice "warm" Schuum.

"Place Your Thumb on it!"

With his cow eyes the faithful steward looks at us, not quite sure if we were serious or joking.

Then he gazes at his black, really black torpedo boatman thumb, and the sparkling wine continues gushing forth beneath him.

"For our loved ones at home!"

"Well, Ensign Berger, for whom have you been drinking?"

"Once again I have had a bunch of luck, Herr Lieutenant; nobody has written to me. Now at least I do not need to answer. "

Here comes the goose!

Dumbfounded we look around, then roars of laughter!

Half a goose! Cross-cut! The front half!

"The cook is to report!"

A Hamburger, "a mound of slime"[102] his character.

"Man, what have you done with the eagle?"

---

[101] Slang for Schaumwein, i.e. sparkling wine.

[102] Schlickschuwer in the original which is a humorous creation of "Schlickschlucker", the mariner's mocking term for landsmen, meaning "mud swallower".

"In the whole half-flotilla there is only one pan in the budget, which No. 3 has it onboard. So now I have to cook it in two pieces. "

"By God and the Imperial Navy exactly nothing is impossible," as my old midshipmen petty officer used to say.

But in all seriousness, the fact is, to us husbands our wives say and do the same using our first names. After all half a goose tastes unequally better than the whole because it is not so fat.

"Guard boat, pick up orders," a signalman bursts into the mess.

Right, today we are also the guard boat!

Our peace is gone!

Christmas Eve and Christmas mood are over.

"Boatswain's Mate of the watch, make sure that all the Christmas trees are immediately thrown overboard! Extinguish all open lights!"

Farewell, Christ child, with your wistful thoughts of home! German sailor, awaken again to warlike action!

"Weigh Anchor! Letter bag clear! "

Semaphore signal from the leader boat "Collect orders from the F.d.T." [103](Leader of the torpedo boat forces)!

---

[103] F.T.D. Führer des Torpedoboots in German. At this date Kapitän zur See und Kommodore Johannes Hartog holds the position.

"If all is alright," my watch officer says softly to himself.

Right he may already have. He is a stern gentleman, the F.d.T. And if he sees something about Christmas, then it will certainly fare poorly for us.

Just all in good time. At moment of weighing anchor everyday life begins again.

And woe to the crew of a boat, when coming alongside his flagship, who makes an un-military impression!

So prevention.

"Boatswain's Mate of the watch, everyone below deck! Only the watch stays above deck! "

"Tell the engine room everyone is to remain below deck!"

Today Christmas Day is for my brave men but let nobody's "father sailor"[104] wander.

"It was all clear," although the F.d T. personally came out when I was lying alongside.

"Signal your flotilla commander, I allow the Flotilla to wish a Merry Christmas!"

"Very good, Herr Commodore! Thank you humbly! "

At least that was the proof that my boat had taken off the foolish Christmas festival and it was no longer noticeable.

---

[104] Vater Seeman in German. Nickname for sailors in detention or under arrest as Father Sailor is a laconic expression for the Rest Home for Sailors.

Now the official letters which I had received for the various flotillas and half flotillas were, "to be deployed", i.e. one approaches from the rear, depending on the wind and wave, from 2 to 15 meters to the receiving boat, and then throws over a sandbag using a long line to which the letter bag is tethered.

The first throw is always 1 meter too short - it's never been different -. So after the second throw the waterproof rubber bag is brought over with its contents. A second line is held by us to retrieve it again.

After an hour we happily finished, just in time to go with the half-flotilla on outpost.

The night passed, the first Christmas Day dawned, a little hazy, but brightly cloudless. The English had not appeared.

Strange! Have we been so very wrong?

“1.Half Flotilla complete coaling!"

So we moor at a collier for a long time.

And now our Christmas service begins, somewhat pagan indeed, if it was compared to sacrificing coals to the hecatombs of Mars.

Every man must come and help as there are no excuses. In the collier they say, the dust in the hold when shoveling coal into baskets is such that one cannot see their neighbors. Then, the basket contents are poured onto the boat deck with derricks and the coal shoveled through circular holes into the bunker.

In and of itself there is not much support for it regarding Europeans. The colored auxiliary people of our opponents are definitely more civilized than us when we coal.

Maybe they are quite right with their "culture".

"What is that?"

On Holy Christmas day, a church service signal shot?

Another burst!

Nearby, so far upstream?

"There he is flying!"

An enemy aircraft on the Jade!

Wonder if he harbored doubts about the location of our fleet?

What drove him to us could indeed be the same.

Anyway, with his amazing effrontery he offered us a magnificent spectacle.

A windless, cloudless winter day. High above the plane and around him little clouds of death and doom bringing shrapnel clouds.

"And he flew away!"[105]

"Do you know the story of the flying Adam? No? So listen! - Service on board. An officer reads aloud from

---

[105] The British launched a Christmas Day air raid against the Zeppelin sheds at Cuxhaven using three seaplane carriers dispatching seven of nine seaplanes. The poor weather made accurate aerial navigation impossible and the aircraft sighted by Callisen was probably piloted by Flight Sub-lieutenant Blackburn who dropped two bombs over a Wilhelmshaven battery which was firing hotly at him. The Zeppelin sheds 35 miles distant in Cuxhaven escaped damage.

the Bible sermon book. In reading he reaches the bottom of the page and has already taken hold of if it with his right hand to fold it to over to the next page. "Adam sits down on a stone and ... flew away."

Scrolls back to previous page. Head shaking. Anyway, two page leaves were joined together! So once again, “Adam sat down on a stone and ...” he desperately rubs the leaves together to separate the sticking pages "flew away"!

Increased head shaking. Goes back to previous page.

"Adam sat down on a stone and ..." - increased rubbing with no success - "and still he flew away."

Our Christmas this year was over.

Mines All Around!

"Herr Lieutenant, we just happen to have a floating mine off the port side."

With these words came the H.I., who was embarked on my boat, onto the bridge.

"Hard to port!"

True, there swam one of these friendly sea mines with the innocuous-looking horn on top!

So an English mine!
There another, there as well!

We were right in the middle amongst floating, hopefully floating mines, twelve in number.

Hopefully floating mines! According to international regulations from before the war a mine is intended to be automatically deactivated as soon as the anchor cable breaks. While there is much in what was written, almost everything in such agreements, which was overturned by our opponents. But it was a slightly relaxing, and just self deception knowing these were floating mines.

Thus initially two boats were connected by a long line to determine whether the anchor cable still existed. The line was swept around under the mine without the latter moving. So it was floating! Thank God!

However we were sensing something sinister, especially since the half-flotilla had to continue on and I was left alone.

Message: "Shoot at the numerous floating mines; your boat to remain here!"

"Forward gunners and machine gunners to the guns!"

Then, at 50 meters closer to the first mine.

"Will you yourselves also blow up?"

We have always been lucky! Why not this time?

The mine is abeam, the boat's nose into the wind, so that exploding splinters do not fly on deck and endanger the crew. In addition, the ship is still. Hitting such a small, bobbing target with a gun from a lurching boat involves far too much luck.

After a few shots the mine is riddled with exploding fragments and sinks.

A few shots for the second mine because it receives a direct hit, raising in the air a huge column of black smoke about 150 meters high and places on deck two hand-sized souvenir fragments of the mine. So carry-on for more!

After all, my shells appeared as somewhat extravagant. So I henceforth allowed only the machine guns to fire. Thus there were certainly more shots needed, but the cost upon hitting was not so high.

One could clearly see the impacts with the binoculars. However, when the bullets struck at an angle to the surface of the mine housing, they left behind only a bright point of impact. And that was naturally more frequent with spherical mines.

But in the case of perpendicular impacts they penetrated through, and then water leaked in every time the sea was running until the mine slowly sank.

Still others detonated a mine. The small machine gun bullet had probably just hit the ignition fuse and caused the charge to detonate.

I have often shot down mines later in the Baltic Sea. However, only one blew up sky high, also by machine-gun fire, a Russian mine. One of the lead caps was hit.[106] The mine had therefore not deactivated when the anchor cable was broken loose, or - it was deliberately intended to float into our waters using the ocean currents, so that a ship is struck at night and receives a deathblow.

This one at least had failed in its purpose. [107]

---

[106] The classic visible protruding trigger horns which has come to exemplify the popular conception of a naval sea mine.

[107] Ironically at this time, 1914, British Spherical Sea Mines were presumed to be capable by the Germans until experience proved them be rather unreliable. British mines improved in 1917 after copying of a captured German mine.

Rammed

The North Sea was glassy. A magnificent April evening, as this moody spring month is so fond of producing, then suddenly brings a change in the weather.

Water vapor was already in the air. Sailors "smelled the fog".

As soon as it is dark, a wireless telegraphist signalman comes from the radio shack with a promising hot signal.

So to bring about an end to our four man Skat that we had to anchor the evening.

"Immediate departure! Details to follow! "

This means to quickly raise anchor. Nothing is more unpleasant than when the leader boat, with the boss on board is ready for sea sooner than others, then drives off without any lights. The accompanying ant swarm of simultaneously outgoing boats at night trying to find the right leader, is quite difficult and often unsuccessful.

"Clear the forward deck!" A necessary command, so that no personnel are at risk in collisions that may occur there.

After we five of the 1st Half Flotilla were together, we went out at high speed. The watch officer was in charge, the commander could therefore go easy for a little while and simply enjoy the evening because an encounter with the enemy was excluded - the night would already be long enough.

About 11 o'clock we were suddenly in a fogbank which was cut off on the water.

Well stay on it!

Only do not only lose the leading man!

The half-flotilla pushes along as a wedge. Just 10 meters from the stern of the man in front and bow of the man astern, often, very often even less. Only not more - then he is gone - and impossible to find.

There is nothing to be seen of the man astern; the fog combined with the pitch-black night, the view from the bridge did not penetrate through up to him, even though there are only 50 to 60 m.

Dead silence on deck and on the bridge, there is heard no signal from the leader. Forward on the forecastle is a petty officer with a large megaphone to obtain oral commands straight from the leading man.

At 11 clock the boss decides to turn around, executing the command issued in such foggy weather is still impossible.

Carefully reverse is ordered! Only a little rudder, lest the rear boats drive into the front ones.

Astern one is happy if the intent is detected in time and so able to keep his boat in formation.

As soon as we have rotated about eight points (equal to 90 degrees) a dark shadow appears on the port side in front of the bow.

"Both engines full speed astern!"

The command is executed immediately within a split second.

I am almost sorry to have given it. Actually, it can only be an enemy destroyer! Although perhaps one of us old commanders lost his formation, and now driving into our turn? Or is it? - Now the outline is visible and actually it's our No. 4.!

At high speed he tries to push past, and then he recognizes the danger to himself. Just do not ram us! This will give us a considerable hole and can jeopardize buoyancy! On the other hand ramming his stern will badly warp it, but can hardly create mortal damage.

It is in vain.

At the level of his mainmast I hit him in a mighty crash. Almost perpendicular I drive into him.

My "snout" is hugely crooked, rolled up five meters to the side. The boat lists, and many of the small forward compartments are filled up.

A second, and then my neighbor no. 4 sits across my bow.

Another crash!

Then again dead silence all around.

Of the four other boats there is fortuitously nothing to be seen.

From the ruins of my forecastle there are soft whimpering sounds.

"Herr Lieutenant, two men sitting forward a trapped, one in the stoker quarters and the other in sailor quarters."

"Break them loose with a crowbar before the compartments fill!"

Sinewy armed sailors and helpful comrade hands indeed bring out both men.

And what about them?

Strong abrasions! Nothing more; in spite of all the thousands of fragments and debris in the front compartments. A lesson for their negligence.

But all 'through the years it had gone well, why should anything happen to us right now? Which is why rather than let the warm soft berth forward untouched and instead coil up in an uncomfortable corner? Especially since it was so hideously cold and damp at the 10 o'clock relief!

A strong dressing-down on top of that was certainly unexpected, but it was a healing balm for their wounds.

My boat had acquired a novel upright floating position. No immediate danger existed. However, No. 4, who was rammed! Poor Guy! It proceeded on in such a pitiful manner! However, there was not much time for thinking. I had had substantial work to do. I tried to steer by the stern, in order to relieve water pressure off the damaged forward hull.

A vain wasted effort!

The bent "snout" functioned much as a strong rudder that the effect on forward steering using the actual rudder was far exaggerated.

It was no good and I had to steer by the bow and head home at the proud speed of 5 knots.

Positive outlook for the day! If it cleared up it could not be more convenient for enemy submarines. -

Then – a fog horn sounds!

"The leader boat makes a position signal!"

All answer in the prescribed order to show that they are still within earshot.

No. 4 is silent.

Who now standing did not still their heart?

Message from the radio shack, "No. 4 is called. But is not responding. "

There, a searchlight flashes faintly through the fog.

It is the leader boat who wants to know our boat number.

"Do You Need Help?"

"No."

Radio message: "The two undamaged boats will return!"

Despite everything, I was immensely glad, at least so far.

And No. 4?

Silently sink on such a night! It will be unlikely to find any swimmers.

The smoother our journey, the more our thoughts wandered to others.

Two o'clock un the morning. Message from the radio shack: "Positioned here and here, traveling 9 knots. No. 4."

"Is communication obstructed?"

Obstructed! The leader boat has to repeat the radio message.

Thank God! A burden fell from our hearts.

Well all those damaged were the same all right!

Towards morning the weather cleared up. One after the other boats came into view. Both healthy ones circled us like a worried mother hen to her chicks.

We fortunately carried on and in the afternoon, paraded among some big ships whose crew curiously considered and criticized our wounds.

"Are people dead?"

"No, on you?"

"No!"

"In the third?"

"Neither!"

Then everything is fine! In battle they may fall, which is inevitable. But only not in such stupid accidents.

The yard will mend the holes again and iron out our snout!

There were, however, associated wrinkles.

My neighbor could compete with me regarding the damaged stem.

No. 4 looked incredible, simply incredible.

I was driven in at the stern, luckily the outer skin is not pierced, but a bulge pushed in from the mast to the stern. All the more respect for German material and German work.

Everything located against the inside of the hull was pushed 1 meter into the boat. All external "ornaments" together with the mast were gone. Instead, I had an anchor placed onto the quarterdeck, and celebrated a happy reunion with my no. 1. (The oldest seagoing petty officer)

My neighbor was staved-in under the bridge. The front boiler room was filled up; the bridge was hanging sadly and had brought down all the means to signal.

And that was the reason for his subsequent silence. Also nothing was missed. Siren and searchlights were damaged; the radio telegraph wireless antenna mast came down.

After painstakingly work a reserve antenna was finally deployed, and then the positive message could be transmitted.

We warmly shook our three unlucky hands.[108] Nor started any grilling nor engaged in retrospective observations. The inevitable paperwork to come would soon enough spoil the good mood ....[109]

---

[108] The 1st Torpedoboat Flotilla War Diary indicates V188, V189 and G197 collided on April 12 requiring dispatch to Kiel for repairs.

[109] The perils to navigation in limited visibility were manifested in the same month as Callisen's flotilla collision when on April 21, 1915 the cruiser Hamburg rammed and sank torpedo boat S21 at the mouth of the Weser. Within the 1st Torpedo boat Flotilla there was a collision between G193 and G195 on February 17 and between V191 and G196 on February 22, 1915.

Navy Pilot

“2. Flotilla to search the sea area west Helgoland for aircraft no. 263 and 264!"[110]

The poor guys! So far they had such 'outrageous luck with their reckless sorties, our planes in the German Bight.

But today! The wind freshens up more and more.

A depressed mood prevails on Helgoland, as occurs everywhere always, when the first losses occur. After that you get used to it, like everything.

Towards morning the search flotilla is back without having achieved anything. All too easily explained by the worsening sea.

And to resolve all doubt, a boat has fished a float. The number is correct. The aircraft are thus destroyed; the four airmen have a found a seaman’s death.

Nevertheless: "7. Flotilla depart, return at noon!"

But they return home without success, as was to be expected.

But all measures must be attempted. The flyers should not and must not lose confidence in their leaders.

At Noon: "1. Flotilla depart!"

---

[110] The serial nos. 263 and 264 correspond to Hansa-Brandenburg W seaplane in use by the German Naval Air Service in 1914 assigned to II. Seefliegerabteilung at the Heligoland seaplane base.

The sea had subsided somewhat, but was still so rough that every minute threw up fresh splashes on the bridge and made the binoculars unusable.

Over and over again the leather cloth had to be brought out to wipe the lenses.

Hour after hour trickles away.

The watch officer announces: "An airplane bears 2 points to port."

"Both engines full speed ahead!"

"Signalman, contact the airplane with the searchlight, so they know that we see them."

Has my neighbor seen it too and wants to have the pleasure to be first? It would be expected, because he has the aircraft right ahead.

Exactly so! He also goes to high speed. Anyway, I must give him precedence; he has the closer to travel.

Especially since - the airplane shoots a star signal! They have seen us.

Well no doubt how they might now be feeling over there?

We approach closer. Everyone watches through the binoculars. Nobody believes his eyes.

"How many people do you count?"

Hesitantly, comes the answer: "Four."

Yes, I cannot help myself, I see four men.

Lord God, indeed we have all of them!

After 36 hours! On an airplane, all four, so how did this happen?

Ensign Stinsky[111], you will not be angry if I mention your name.

He had seen the other aircraft going down onto the water, crash into the rough sea and how the two occupants were fighting with the waves.

Without hesitation he goes down and after a lot of effort takes on the swimmers.

But, alas! The load is too heavy. The engine is not its equal.

Go home riding on the water! Certainly, if one could! But the sea rolled in from the West, and the aircraft would inevitably capsize in such a sea.

Oh when, is the flotilla to be looking for us, just be pretty economical with the signal cartridges! In the next six hours it serves no purpose to consider which to burn, there can still be no-one in the vicinity.

So Stinsky stays at the controls, the other three squeeze together in the observer's seat in order to warm each other in their sodden clothes.

The day passes.

The night comes, eerie dark. And the white heads on the sea are becoming more numerous, the mountainous waves higher and higher.

---

[111] The only Stinksy in the German Naval Air Service is Max Stinksky(1895-1918), later when holding Lieutenant rank the celebrated leader of I Torpedostaffel operating out of Zeebrugge in 1918.

With superhuman effort Stinsky holds the plane towards the sea; he can always start the engine when the airplane wants to capsize.

Is it true? The fuel is running low?

What now?

A few expendable items are quickly tied together on a rope and set over the side as a sea anchor side. Now the aircraft, lying-high on the water, floats swifter than the sea anchor, and then is stable in a seaway.

A jolt! The sea anchor has broken! With difficulty the airplane with the last remnant of fuel is brought back to maintain the seas.

In a hurry, a new sea anchor is manufactured from guns, ammunition and whatever else exists. Only a pistol is kept back with four cartridges.

The last chunk of the scanty supply - for they had flown in additional two hours - was distributed fairly.

For twenty-four hours they drift - no one comes.

"Surely they have found our floats and therefore continued searching on" expressed one of those concerned.

The day passes. One is faint from overexertion and starvation. A float is leaking, so that one must constantly maintain equilibrium weight on the sound float.

Four clock in the afternoon. "Star Shell![112]"

---

[112] An artillery illumination projectile.

Now he was lively again, no longer lying lifelessly in the seat!

"A submarine! That is all I need! Now into English captivity!"

The ship is large, next another one, then another! Hooray, a German flotilla!

When lifting the aircraft the battered leak proof float collapsed. It had outlived its usefulness.

V188

The Ems

"1. Half Flotilla to temporarily assume security!"[113]
Very Nice!

Exactly four weeks we had spent on the Ems and were thoroughly acquainted with its many surprises. For three days I had been all this time in port. But the sailor does say:

"What do I need to go ashore; because I can see the country from on board."

In black and white we had it, too: Three days at sea, two in the harbor.

But Mars like Neptune cannot seem to read; or we had failed to provide them with a copy.

Enough, it was, as always in the Navy, once again different.

We had just moored in Emden's outer harbor, when the order came to sail.

What did it matter? Out we had to go! Who would have guessed that it would be the last trip for my brave boat!

It was a gorgeous summer day. The sea glassy; a hot burning sun drove everyone off duty, up on the tarred deck.

---

[113] V188 and the remainder of the 1st Half-Flotilla were ordered to provide security for a minesweeping unit. See Appendix D&E at the end of this book.

In addition, it was precisely the birthday of the start of the war, "our" war's outbreak.[114]

Exactly one year ago, we went to sea to look at the flag of the French Republic's President. There was some exchange of words. Memories were refreshed and probably the newcomers here and there turned up some, good old seaman's dish.

Above us crossed an airship, exchanging a cheerful semaphore with us.

Signal from the leader boat: "Detach! V188 northern wing boat. "

During the signal's descent we parted from the leader boat, and at high speed it went alone to the northwest.

Just before I was in position, a Dutch trawler came into view, I head to examine.

Herrmann, who seemed particularly suitable as a Hapag[115]-officer to check the papers, was sent over in the dinghy while we created our circles.

Meanwhile, the weather gods had his friendly face badly warped and sent us a nasty thunderstorm forecast.

"What to one is his owl, is to the other his nightingale!"[116] So we officers stood on the bridge and

[114] The commencement date for this chapter is July 26, 1915. The first chapter of this book dealing the outbreak of war is dated as commencing on July 26, 1914. The date is presumably a literary device to connect the events contained therein. Germany was still neutral on July 26, 1914.

[115] Hapag, acronym for Hamburg-Amerikanische Packetfahrt-Actien-Gesellschaft. A German merchant shipping line.

[116] Platt German saying for: Everyone sees things from his perspective!

were happy as a drenched Herrmann returned with a sweet and sour expression. After all he did not need to provide the mockery.

On the steamer everything was found in order.

So onto position!

The officers went to lunch in the mess and left me, as always when at sea, back on the bridge. For the commander there is no relief.

Seven minutes later comes the cry from the forward mast crow's nest:

"Submarine to starboard!"

"Hard to starboard!" Both engines full speed ahead! Run over him! "

At the moment I jump over to the starboard side of the bridge, and already see the torpedo discharge splash next to the periscope, about 100 meters from us.

Perfectly straight is the bubble path of the torpedo coming towards us.

It will hit! As there is no escape. After five seconds, he will have us. All too often we had seen the same picture at peacetime exercises and rejoiced at the beautiful shot.

Even I jump to the alarm bell sounding in all parts of the boat and calling the crew to the guns.

On the bridge, all observed almost peacefully the bubble path, more curious than worried whether he will probably hit.

It must indeed pass among us. It would be an almost incredible luck for the submarine if the torpedo would travel to the right depth in the short firing range to strike our shallow moving boat.

The first bubbles reach the ship's side.

Even I want to breathe again, believing the torpedo ran too deep, because it has not yet detonated.

A sharp blow occurs. The whole boat trembles. A 100-meter high column of black smoke rises next to the officers' mess and snaps the mast with it.

Countless debris and fragments are hurled on the boat. A piece also finds its way to my lip to always remind me of this fateful hour.

My poor proud boat!

On one side the stern hung still attached to the taller foredeck, but heeled over almost 30 degrees. Deeply it lay in the water with its wound. Movement was not possible between the two parts of the vessel.

And my three officers with whom I spent almost a year in such a small space had found a quick soldier's death. Two comrades remained in death with them.[117]

But my boat floated, and it was necessary to save it!

---

[117] There were five fatalities in the torpedoing of V188. The date of torpedoing is July 26, 1915. Approximate local time is 12:36 p.m. Events are detailed in Appendix F, Callisen's Loss Report.

Meanwhile, the submarine drove around us and looked at from time to time through the periscope at its glory and our impotent rage.

A tremendous roar of the steam flows out of the rear engine pipelines whose crews were saved by a miracle. It is quite impossible for voices to register over 1 meter away.

One after another I reach out and shout my orders in his ear:

"Helmsman, with the forward gun constantly bombard the periscope to prevent the submarine at a second shot!"

"Launch the dinghy to pick up men from the stern!"

"Deploy the reserve F. T.! "[118]

"Support the bulkhead between the two engine rooms!"

"Contact the neighbor with searchlight!"

Answer: "Searchlights have failed!"

That's the limit. The wireless telegraph had "come down from above" with the mast. Now I had no means to reach the 6 nautical mile distance to my neighbor.

And he had seen nothing, because a thunderstorm gust had been between us.

"Fire forward gun with the highest elevation at the neighboring boat!"

---

118 F.T.= wireless transmitter.

He had yet to see my shell splashes!

Finally! Six long minutes after detonation he turned towards us! The commander told me later that he thought I was in battle with destroyers who had also fired on him.[119]

"Front gun shoot at the periscope!"

It was my only signaling means.

Had the other boat coming seen our opponent in order not to be torpedoed as it was, of course, his is intention?[120]

It almost seemed as if fate also wanted to prevent this as well.

The submarine was now gone to the other side and waited with extended periscope for my neighbor who arrived just astern of me. With strenuous effort the gun crew sought to turn around the gun barrel to port and onto the submarine, but my boat was listing over

---

[119] Callisen omits details, certainly due to wartime security, regarding the action. See Appendix D with an extract from the Official History of the War at Sea regarding V186 attempt to torpedo the British submarine.

[120] E.16, (Commander C.P. Talbot) the British submarine in question actually did launch torpedoes at the German torpedo boats coming to the rescue. Citing from the Naval Staff Monograph Volume XIV, Home Water Part V, p.58 "Returning to his station, he was in 54.11 N.5.23E, at 11.20 a.m. July 26, when he sighted three German destroyers apparently hunting on various courses. At 12.36 p.m. he got a good shot at one of them, hitting her and blowing off her stern. The two others came up at full speed; he fired a torpedo at each of them, but in both cases the destroyers avoided them and continued their attempts to ram E.16. Commander Talbot retired to the northeastward, watched by a Zeppelin, which kept him diving till 7 p.m."
E.16 had arrived in the area the on July 25 and experienced an eventful 24hours: chasing a U-Boat, attacked by a seaplane, twice by zeppelins, hunted by five destroyers and investing several hours submerged in freeing itself from a fouled heavy net. See Appendix G to view E16 log of events.

so heavily that the slewing mechanism failed. But finally succeeding using a wooden lever, and already has the next shot ready to show our savior the location of the submarine.

It was high time! Hardly had hit our shell hit, and already the torpedo struck the other boat.

We all probably stood with still hearts for a moment.

Thank God! A detonation failed to materialize, so the torpedo boat evasion must have succeeded. And so it was.

However, my brave boat that had carried us for twenty minutes offset this latest incident with its deathblow. A stern seaway made by the other boat's high speed, broke apart the two parts.

The stern sank; the ship's forward section slowly listed about 45 degrees to port.

"Don life jackets!"

Like a swarm of buzzing bees the men chaotically collect each vest bearing his number. But soon order returned. As was so often practiced, the crew went on deck; only the gunners and the boiler room personal remained on station.

"Evacuate the boiler and engine room!"

The command is repeated in all compartments.

Then they climb up the narrow stairways of the boiler rooms, where they have endured so well almost to capsizing.

And now comes a moment, as I have never seen him more beautiful.

"Three cheers for His Majesty the Emperor!"

Powerful and passionate, as it has never been heard from sailors throats, caps were never swung higher! And a nicer echo has never been answered in a mountain forest as the concurring shout of the crew of the other boat! -

"Hands up whoever has no life jacket!"

No arm rises.

"All hands over the side on the starboard side! Swim away as far as possible! "

A sailor comes to me with slung folios.

"Shall I take the signal book?"

"Since you will probably drown! Throw it into the dinghy! "

The thick book with its heavy lead covers has a rather considerable weight. But it must never fall into enemy hands under any circumstances.

The boat lists precariously to one side and is close to capsizing.

As I stood alone on the bridge, again three cheers ring out from the water for our brave boat and then to the commander. One should not interpret my mentioning as vanity, I owe my crew their just due as a testament to their magnificent spirit.

With a heavy heart I then leave my boat, which has so faithfully served for almost four years.

Proud was his life. Proud was his death.

Once again, it reared up in impotent rage; high it stood upright in the water, then slowly descended forward bolt upright.

Again three cheers rang from the water. The floating crew made it to the sinking boat and brought off our remaining comrades.

The masses of almost ninety swimmers proved difficult for the other boat. The only remaining lifeboat took just five men, four wounded and a rower.

Taking us directly was impossible because of the proximity of the submarine, since this would have inevitably destroyed the stationary boat. Therefore, the cutter launched from far away. After taking in twelve to fifteen men he then rowed back to set them down at another place with constant change of its human cargo onto the torpedo boat.

The outdoor pool party accordingly lasted quite a long time.

Again, I cannot praise enough the behavior of my wonderful crew.

"The commander first!" And it is repeated by many voices: Is loudly called when the cutter draws near.

Downright unparliamentarily expressions I had to use to get people in front of me into the boat.

Next to me appears a swimmer in his virgin birthday suit. Coming close it turns out to be the ensign from the leader boat.

"Can Herr Lieutenant still swim?"

"You are probably relying on God! What do you want here? "

"I just wanted to make room in the boat, so that one more can enter into it."

"Now hurry back inside! We have on lifejackets."

Finally, the last man was recovered.

The torpedo boat took us to Emden, where a warm welcome was given to us.

After fourteen days we were united aboard another boat.[121] In good spirits and ready for new deeds.

But the English thought we were done, who as "gentlemen" even pay for our expensive death. In "Lloyd's List"[122] it was in black and white, to read that the commander and crew of the British submarine "E 16" received the sum of 465 pounds, in a London Prize Court judgment for the for the destruction of our boat.[123]

Let them receive their well earned blood money! We grant it to them!

---

[121] Callisen assumes command of V-191 within the same flotilla.

[122] The famed British shipping journal.

[123] London Gazette #29687 page 7478 Announcement dated July 28, 1916. Prize money was 5GBP per head or 93 crew complement aboard V188. E16 failed to return from a North Sea patrol in 1916 after striking a mine.

Windau[124]

"How did you come to the northern Baltic Sea?"

"Very easy and very fast. A radio message: First Flotilla proceed immediately to Libau".[125]

Five minutes later, the anchors were cleared, and in good spirits we were away, I personally have not yet been up there.

But the joy was short-lived.

The Baltic Sea looks so pretty small on the map, but it has introduced to us hellishly uncomfortable waters.

From Libau we were each sent away for trade protection. I have since thoroughly revamped my views on the small distances in the Baltic.

The bright spots on these patrols were the short hours of rest in Windau at the mouth of the same river in Kurland, a not unimportant export port for timber and grain. Before the war rather extensive wharves were created, and far out to sea a lengthy mole closed in a semicircle harbor basin. Only a small opening was left for the river's water flow and at the same time an entrance for ships.

But like everything in Russia, this was also imperfect.

---

[124] Windau in the original German, Ventspils is the contemporary name in Lativian.

[125] Libau former Russian naval base captured by the Germans in August 1914 and used as a staging base for operations into the Northern Baltic. The flotilla received orders to depart to Libau on August 11 1915.

A narrow shipping channel went from the outer entrance through the basin to the river to the quay areas. All the remaining parts were unusable even for shallow-draft vessels.

Now the harbor looked desolate.

Before their departure the Russians had demolished everything that was worth destroying.

The lighthouses were blasted into rubble, in the outer breakwater large breaches were opened to the sea which should complete their destruction.

The river mouth was blocked by four steamers occupying, transversely to the current, the space between two piers. Only railing, funnels and masts were sticking up out of the water. Obliquely they lay on top of each other as if the innermost steamer had fallen and the crushed the others under itself.

At this sight, I could not help thinking of that very tall lieutenant who appears for the first time before the front. In the second section a man grins.

"Damn it, why are you grinning at?"

Awkward silence.

"Out with it!"

"I thought if the Lieutenant should instantly fall, we will all tumble over."

Today, of course, these ships were no longer an obstacle.

Good heavens, but how it looked from inside the river!

Instead of the quay areas here and there charred remains of beams and piles – the kerosene had apparently not been spared in kindling the fire.

The powerful elevators of a huge storehouse sadly hung down buckled. The legs were dynamited.

“Does nobody then come to raise it again?”

No, it can wait."

The city has not suffered from the fire. I believe only a lumberyard and four to five houses had.

But nobody is the city. On the road you meet no one, apart from German Army Reserve soldiers. The houses are barren and abandoned.

Close to river on a hill stands an almost unnatural rectangular connected heavy construction: The Old Castle, before the war the Russian courthouse and jail.

"We need to look at it."

A very small door leads us through thick walls to a dark hallway. From here even smaller doors lead to the old living quarters, and last office of the Russian magistrate together with his secretary.

I enter, but immediately recoil.

"This is something I have not seen before!"

"Cheer up," said my companion.

The Russians have been in quite a hurry in their flight.

The floors were a veritable jumble of torn documents, letters, photographs of criminals, books, uniform pieces, rolls, shirts.

Not one piece was intact.

Nevertheless it was as if the departing magistrate had been afraid to hold a match to the venerable castle, lest everything should be destroyed.

Through a dark corridor, we entered a spacious chapel, the only clean, decent room in the whole building.

Here everything was still in order. Countless old magnificent holy pictures hung on the walls, church records bound in pigskin were carefully on their shelves, and only the eternal lamp was extinguished.

Using a chicken stair, as primitive a wood construction as a layman can make, we climbed onto the loft.

Flooring was missing for many years now; pigeons nesting in all corners. Between them in stylish disorder were clergy vestments, processional devices and Saints' pictures, they probably could find no more room in the chapel for them.

"But we still need to see the prison."

"Yes, that is certainly worth seeing."

In inner court, as we find even in old German castles, we meet a Reserve sergeant with a huge bunch of keys.

We ask if we can enter.

"Well yes" replies the worthy Saxon, who is now appointed jailer.

A narrow staircase leads to the first floor gallery, which is set up as a prisoners' workshop. A pity about the beautiful archway.

On the second floor is the prison.

"What it is for now used?"

"There are fifteen civilian prisoners therein. Some convicts, some people in custody for theft and espionage. "

Curious, we step up to the observation holes, which are embedded in the doors.

"Well, that looks very decent in there!"

"I imagined a very different dive in a Russian prison."

The old palace rooms in their original grandeur! Disappointed, we walked back down.

Additionally the sirens sounded, the agreed sign that we should go to sea again, "our" trenches.

Bad Weather

"Well, in my experience I've never seen such bad weather up here!"

"It just proves true again and again the saying: only fools go to sea."

"And many would still spend another thaler[126] if he could ride along," the H.I. injected laboriously out of his tortured throat.

He feels his two hundredweight proportions is not yet quite comfortable in the tight torpedo boat. The quarters are so narrow that in the morning in order to get into his trousers he must stretch his legs out of the door into the mess, where the other gentlemen are sitting at breakfast. As a "Bather" he can afford to get up last.

Maybe he succeeds in becoming slimmer. Since this is undoubtedly the purpose of his duty assignment to a torpedo boat.

And he boarded just before our trip Baltic Sea. The torpedo boat with its continuous vibration works much better than electric vibrating chairs. Seasickness is inevitable given the bad weather and salty bread.

In fact having to travel the northern Baltic Sea in December is something other than beautiful.

---

[126] A thaler is any of various silver coins that served as a unit of currency in some Germanic countries between the 15th and 19th centuries.

We stood on the bridge muffled as at the North Pole. Several layers of wool over the other, oilskin, cupids and a towel wrapped around ones neck so that no water runs down.

For now the sea is still before us; but how will it be on the return journey if we have to drive in against wind and waves?

"Dear God, let morning be, evening is by itself.[127] At least for us torpedo boat men. "

The wind velocity is alarming.

Radio message: "Torpedo boats are to seek protection among land."

Thank God. But it can still take four to five hours until we get there.

The leader boat continues onward.

After half an hour there seems to me to be something wrong with this business.

"Megaphone"

"Good close approach".

10 meters between us.

"Leader boat! Did you receive the radio message? "

"No! Repeat! "

Finally, he turns around.

Splash! The first sea inundates the bridge!

---

[127] A rendition of the German proverb Dear God, let evening begin, morning will then follow on its own.

That's what happens when you do not watch out. Like a drowned rat I shake off the water, but without success.

The oilskin immediately ices up in the extreme cold. It becomes covered with ice crust and is hard and uncomfortable. If I move, it cracks.

Especially when it comes in contact against the sea.

It is now lively onboard. Every few seconds the bow rams down deep, scoops back water and throws it back over the bridge.

"Front man out of view," reports the helmsman.

The panes of glass in front of him are icy, and the front man sails dimmed without any light. As always, in order not to betray to the enemy.

"Open window panes!"

At that moment the wind piercingly drives in water spray and covers everything with an ice crust.

"Signal man! Morse this over there, please employ stern light! "

The poor guy.

In order to send his Morse message with a small electric lamp, he must climb over the protective bulwark and now gets every wave on himself.

The rear one supports him.

The leader boat understands and employs a light.

Anyway it cannot be seen from far away. Certainly it will not betray us.

Up and down the white stern light dances in front of us in regular cycles and disappears behind each wave crest.

Damn, now it's snowing too!

"Helmsman, stay alert!"

The leader boat is gone.

Carefully some headway is increased, and then his stern light reappears, almost next to our bridge.

The sea is still rough.

From the flooded crew quarters, one after another appeared on deck. It is good that it is pitch black night and one does not recognize their facial complexions.

Ah! That feels good, fresh air and a thorough soaking!

But for how long? A "Chain Plug"[128] is inevitable - and in the background smirk the healthy with their happiest, most innocent face.

Well, do you not get used to this type of sea voyage?

No, never on a torpedo boat. Everyone suffers from seasick due to the repeated incredibly violent and irregular movements of the boat. The oldest deck officers, sailing on torpedo boats for twenty years or more, suffer as the youngest recruit.

And yet even ask someone with red seam: "Would you prefer to be on a big ship"

A telling smile is the answer. -

---

[128] Footnote in the original text. A term for extending the anchor chain, used here as an analogy for victims of seasickness.

I did say we needed four to five hours until a sheltered bay? Good heavens!

That night was no question of going inshore. In such weather no one could know where on the dear Baltic we were actually located.

With difficulty, my completely soaked brave orderly arrives on the bridge.

"Shall I bring Herr Lieutenant something to eat?"

"Yes and some tea."

"We cannot cook; the galley fire is also out of service."

"So bread then."

"Ugh, the Devil, there is nothing to eat except salt."

Yes. Below deck everything is under water, and up top you cannot move.

The old song: Once at sea one only has the quiet hope to maybe get a piece of 'dry' bread.

At twelve noon we were in Windau.

Semaphore signal: "Immediately load coal! By 3 o'clock accompany "Bremen"[129] to sea! "

So back into the wet fun!

But being in the company of a big ship has at least the prospect of better navigation, which is sometimes a bit difficult in torpedo boat conditions.

Nevertheless from some of us a rough man's curse slipped out.

---

129 Cruiser Bremen

Worse, it could certainly not be.

Certainly! To err is human - that would be made clear to us once again.

Laboriously we struggled behind the "Bremen" to boot. Our only pleasure was that she danced like we did. Over there they are not enjoying the seaway as well.

The hours crept by painfully slow. I had gone two days and nights without interruption on the bridge, in a continuous seesawing squat to maintain balance on the rolling boat, and clinging with both hands on the railing. Inevitably whoever releases their grip is knocked about the moment a wave washes over. It is impossible to follow the movements of the boat wearing cupids.

Then, at daybreak the next morning, I experienced the most horrible moment of my life.

We ran diagonally in front of the sea. Suddenly a house high breaker, about 2 meters higher than the boat, seizes the stern and breaks over the deck.

Slowly we lean, almost to capsizing under the roar of the breaker and the crash of the splintered mast and the shattered cutter.

My heart stopped.

"Hard to port! Both engines slow ahead! "

Against the seaway and as slowly as possible, so that it rolls away faster than the under the boat.

I did not believe we would straighten up. This includes all people below deck, and those above deck wearing oilskins

At the moment I had in mind the capsizing of "S 26" in the Elbe estuary in 1897 under the command of the Duke of Mecklenburg.[130]

The feeling that all of his people are probably completely lost is probably the worst thing for a commander.

The seconds appear not to want to move forward.

Finally! Slowly, very slowly the boat rises again.

Next to me my old helmsman, a “seafaring man" mumbles:

“Young lad, I would not think or wager on how close over we came.”

At the same moment sounds from aft the cry that places teeth on edge every sailor, even the roughest torpedo boat man: "Man Overboard".

"Hard to starboard! Both engines to make full speed! "

Turn the boat around! Use the shortest possible time to transverse the seaway!

A petty officer arrives on the bridge.

“A man has just gone overboard with the portside machine gun."

"What was he wearing?"

---

[130] On 22 September 1897, S-26 capsized and floundered in a storm with the loss of her commander and seven others.

"Oilskin, sea boots."

So hopelessly lost - he had to sink like a stone.

We slowly headed back into the seaway and used the moments of dryness to look for him.

Hope faded with every passing minute.

The honest fellow!

He wanted to clean his machine gun, and then the sea grabbed him and washed him along with his weapon over board.

With a heavy heart I finally gave up the search.

The first time in my life that a man had fallen overboard on my ship and had not been saved.

Lovely is the responsibility, but also infinitely heavy. -

At noon we reached Libau. I had stood up for fifty-five hours, thrown about irregularly back and forth a second like on a swing and every five to six seconds doused with a bucket of water whose contents immediately froze.

Finger thickness ropes were iced arm thick.

The boats looked as if they had escaped a winter wonderland ....

Sunk along with S. M. S. "Bremen"

"At 3 clock be ready for sea!"[131]

The weather had abated, the snow stopped. There was still a slight swell from the West, the last breaths of an excited sea.

For weeks the leader boat[132] and ourselves had been assigned to the "Bremen".[133] Thus we ran out now once again to common action.[134]

In a circular disc the full moon rose over the Russian lands and quite unnecessarily lit up the cloudless Nordland night.

Our course was to the west.

We were sailing for two hours. I stood in the charthouse on the bridge setting the course on the map.

---

[131] December 17, 1915

[132] V-186 is the leader boat

[133] The 1st Half-Flotilla and the cruiser Bremen had been assigned to patrol the Gotland – Windau corridor and deter or intercept Russian attempts to mine the swept ship channel. A cruiser and torpedo boats guarded the route by day and torpedo boats at night. Both the Russians and Germans relied heavily on mine warfare to constrain and attrition their opponents and protect their own waters. For example in the Baltic Sea, the Russian Navy laid down a total of 38,932 mines in both offensive and defensive mine fields, which cost the Germans 48 warships sunk and 21 warships damaged.

[134] On November 20th, 1915 seven Russian destroyers raided the German patrol line near Windau and sank the patrol ship Norburg. The Bremen and accompanying torpedo boats provided security for the small and lightly armed auxiliary patrol ships in use on the picket line.

Suddenly a dull noise! A shiver! The glass panes splintered all around me![135]

After a split second I did not know anything further.

After a short time I come to my senses and find myself again on the deck in the narrow space between the funnel and chart table. The light of the full moon penetrated in to laugh at me.

On all fours, I crawled out of my cage. The detonation smashed the closed door of the chartroom. Otherwise my boat would have been my prison dragging me down into the depths.

On the starboard side of the bridge the watch officer "Collected" his "back bones together." Quite puzzled he rubbed his elongated backbone that now had a very tangible result of his involuntary set down.

The port side of the bridge was destroyed. Some people had foot and lower leg fractures so typical with underwater detonations. The helmsman, who was under the roof searchlight, climbs down from the top.

"Where did you come from?"

"I do not know anything." He was all confused.

The iron roof above him had opened up quicker than the man could fly and allowed him to pass without injuring him in the least.

The view from over the bridge railing.

---

[135] 5:10pm local time

The fore deck lay deep under water; the boat was hit on the port side near the front boiler room. The crew front living quarters were on top of the forecastle.

"All hands to the stern! Ready the life jackets! "

To the watch officer: "Get all the people aft!"

There was not much time to lose. I waded towards the stern through the water up to the knees.

“Launch both boats!"

"Herr Lieutenant, the cutter is unusable."

A leap and I was there.

Bad luck! "The detonation passed through his cleats and storage!"

"It may have been the great pressure!"

"Who has no life jacket? Hands up! "

All too many hands rose, because the detonation had destroyed all forward life jackets.

"Hands down! Which of you cannot swim? "

There were even more!

"Six men into the dinghy!"

"The rest stood still! Three cheers for His Majesty the Emperor! Hurrah, hurrah, hurrah! "

I'm reach to my head to pivot my cap - it is missing. Well, then it's even without said cap.

Already arrived alongside were the rowing boats of the "Bremen" and my leader boat.

"The wounded and men without life jackets into the boat!"

"Coxswain do not take too many!"

The leader boat itself is approaching from astern to come alongside.

"Swing in the torpedo tubes!"

The torpedo officer is already sitting in on the spoon[136] of the tube and removes the warhead pistol. All that was missing was for our own torpedoes to go off when the other boat approaches!

The electric lights are still burning.

"Evacuate engine rooms!"

Upon doing so, the leader boat has his foredeck at my stern. Too far forward it may not proceed, as my boat can at any moment make a deep plunge if cold water blows up a boiler.

Oh woe! It rolls violently. Climbing over is impossible.

The "Bremen" positions in front of it to create a lee.

Then, quite unexpectedly, for a moment the boats lie calmly, as if God would have that we might be saved.

The men quickly climb over.

"Damn! Is the cold on my head! "

I wrap my scarf around my ears.

The last men climb over.

---

[136] Spoon is the term for the above water torpedo tube fairing which shields the protruding torpedo warhead. It is indeed somewhat spoon shaped.

The rolling starts again.

We four officers are still on the brave boat, which already raises its screws into the air, and cannot abandon it.

"Leader boat, you can cast off! We are staying here and want to try to tow it. "

"Nonsense, I will come alongside, again!" Calls back the H. Leader (Half Flotilla Leader).

The chief engineer reaches into his pocket.

"Damn!" Then he wants to go down into his compartment.

One of the lieutenants sees this and grabs hold of his two hundred pounds, and asks, "What do you want below?"

"That's none of your business!"

"I already know what it is. You will never get back up again!" So saying, he speeds downward and comes up again with a small, shinning object. Silently, he gives it to the H. I.

The leader boat is alongside again.

I am unable to leave so easily; I cannot believe my boat is sinking.

Again, I call over: "Why do we not try to tow it? Apparently it is holding together! "

"No, you must come over!"

My second Lieutenant, a Bavarian, who returned from months of travel as a stowaway from Bangkok, climbs the tower, he is wearing no more than a shred of a coat.

"Stop! What do you want? "

"To take the flag!"

"No, leave it! My boat is expected to sink with the flag waving! This is not a regimental flag! "

He now again remembers his distant recruit period training. He comes down rather crestfallen.

"Quickly over please!"

One of the lieutenants pauses in thought; he probably wants to be the last.

"Next, do not cheat me."

The H.I. appears one foot on the railing and with his hands gripping the bridge of the other boat. All of his bodily abundance hanging in the air.

At this sight a stoker forgets he is a soldier, and says frankly: "That's fine that you are here". He holds him and tries to support him. The HI lets go, and now they both land on deck.

I myself have only one foot on the railing of the other boat as countless men's fists grab me and pull me over.

My beautiful boat I would never again re-enter.

And yet it seems as if it does not settle deeper.

I report to the H. Leader.

"We want to try to take your boat in tow. But I think it is impossible that it will continue to float."

"Cutter alongside!"

A cutter from the "Bremen" is approaching.

"Six volunteers forward!"

My lieutenants are already sitting in the cutter, my sailors step forward without exception. Then quickly select oldest into it and rowed over.

Too late! The H. Leader was right. We are not there yet as the boat rises steeply with the stern high up into the night.

About ten meters she rose above the water and seemed to stand on the bottom. For a moment it struggled there, and then lay over to the side. "Three cheers for our old boat!"

Then it was gone.[137]

With a sad heart we returned to the leader boat.

Hardly have I arrived again on the command bridge of the leader boat, that an immense detonation on "Bremen" startles us.[138] Glowing flames about mast high reached into the sky. No painter could represent it.

Then a black cloud lies over its bow, so that we, who were right ahead of her, saw nothing from the ship.

---

[137] 5:45pm. 35 minutes after contact with the Russian mine.

[138] Bremen detonated two mines at 6:04pm

It catches up to us right now that the whole ship is gone.

No, thank God, she floats, but the forecastle is deep in the water.

She is sinking quickly.

Assistance is urgently needed there.

With high speed, we intend to maneuver alongside aft and take off the men.

Three cheers resound through the night, and then the bow disappears under water. The stern rises up high, black full of people.

We have to move quickly!

With a lurch the beautiful ship disappears, taking the whole crew with her.

Only a cutter is intact, plus a few rafts.

What is not swallowed by the whirlpool is safe from it and then picked up by us. Many we lift directly from the water onto our deck.

But how many brave have been killed in the huge blast and subsequently pulled down into the vortex. And among them my brave crew, unless they managed to survive on the "Bremen"[139].

Further, we were now alone on the sea. When will we suffer the same fate?

---

[139] 38 died from V191's crew. In total 8 officers and 102 other ranks were saved and 11 officers and 287 other ranks were killed including the Bremen's captain.

But our God meant differently with us.[140]

In a few hours we were in Libau, where we found a loving welcome on the hospital ship "Schleswig", which will be familiar to many due its beautiful voyages to the Orient.

Two days later we were on the train to go home.

Many a traveler between Königsberg and Hamburg will shake his head in astonishment when he saw us in our little more than scanty uniforms.

With joy I remember now the somewhat apologetic, partially doubtful concerned looks in the dining car of the express train about my state of mind what with my plastered eye, my torn jacket, patent leather boots, collarless, when I courted the pleasures of the table. And gratefully I remember a fellow traveler who entertained my crew in the dining car in a gracious manner.

---

[140] The Russian destroyers "Novik", "Winner" and "Bully" laid 150 mines on the Spon Bank on December 16, 1915. This is the field which sank V191 and Bremen. The Russian mine laying was hampered by unsure navigation and the novice crews on the Winner and Bully out on their first wartime sortie. The veteran Novik, on its twelfth mine laying sortie, efficiently placed its mines. The commander of the Novik was concerned the uncertain efforts by the other two crews would result in a less dense field than necessary for full lethal effect. The mine laying was completed approximately half past midnight on the 17th. V-186 reported both ships lost due to submarine attack; reporting one torpedo passing 10 meters forward and the other- 20 meters astern. Russian intelligence fostered the deception of a submarine having sunk the ships leaving the minefield intact. On December 23 this minefield also sank patrol ship No.15 (Freya). The torpedo boat S-177 and S-176 rescued the survivors and broadcast a mine warning although S-177 struck a mine exiting the field and sank within a few minutes with the loss of 27 crew. The Germans swept the field in the following days.

So we arrived in Kiel, not by boat, as we had gone out, no, on land! And far too many of my brave men were missing!

It reads short and simple in the newspaper:

"On the afternoon of December 17 fell ..."[141]

And yet, nothing more terrible can be experienced.[142]

[141] An extract of the official announcement. The complete one simply stated "On the afternoon on December 17 HM small cruiser "Bremen" and one of the torpedo boats that accompanied it in the eastern Baltic were sunk by a submarine attack. A significant part of the crew were rescued.
Chief of Naval Staff"

[142] End of original text.

Appendix A
Friedrich von Callisen
Career Information[143]

1883 [May] Born

1901 [Apr 10] Entered Navy

1904 Sea Cadet, Pre-dreadnought Kaiser Karl der Grosse

1904 Lieutenant [Sept]

1905 Lieutenant, unprotected cruiser Falke

1906 Sr. Lieutenant [June], unprotected cruiser Falke

1907 Sr. Lieutenant, I Shipyard Division (Kiel)

1908 Sr. Lieutenant, Torpedo Boat G.137, Training Flotilla

1909 Sr. Lieutenant, Torpedo Boat G.137, Maneuver Flotilla

1910 Sr. Lieutenant, Pre-dreadnought Württemberg, (as an instructor)

1911 Sr. Lieutenant, Pre-dreadnought Württemberg, (as an instructor)

1912 Captain-Lieutenant, Commander of torpedo boatV188, I Half-Flotilla, I Torpedo Boat Flotilla.

---

[143] References utilized include the Rangliste for 1904-1914 and the postwar Ehrenrangliste.

1913 Captain-Lieutenant, Commander of torpedo boatV188, I Half-Flotilla, I Torpedo Boat Flotilla.

1914 Captain-Lieutenant, Commander of Torpedo Boat V188, I Half-Flotilla, I Torpedo Boat Flotilla.

1915 Captain-Lieutenant, Commander of Torpedo Boat V188, I Half-Flotilla, I Torpedo Boat Flotilla to July 26

1915 Captain-Lieutenant, Commander of Torpedo Boat V191 to December 17

1915 Captain-Lieutenant, Commander of torpedo boat S50

1915 Captain-Lieutenant, [Dec] Assigned to the Torpedo Acceptance Commission

1917 Captain-Lieutenant, [Oct] Commander 10th Torpedo Boat Half-Flotilla within the 5th Torpedo Boat Flotilla

1918 Captain-Lieutenant, [Jan] Commander Ottoman Torpedo Boat Flotilla (Bosporus)

1919 [Nov] Departed Navy

1968 Deceased

Appendix B
V188 at the Battle of Heligoland Bight, August 28, 1914 as narrated in the German Official History of the War at Sea.[144]

The next adjacent boat on the left of the flotilla leader[145] was V-188. At 8:20 a.m. the latter sighted a submarine on the surface bearing west-southwest about 8750 yards distant. The Captain, Lieutenant Callisen, headed at once toward the boat at full power and, when recognition signal was not answered, opened fire at 6500 yards. However, even before the second shot was fired, the hostile boat was almost completely submerged and only twice was the periscope sighted for a short time. As the radio transmitter on V-188 was out of commission, a report thereon could unfortunately not be made. However, the following information was received from G-194: "Four enemy destroyers 142 epsilon[146], heading south." The Captain at once appreciated the danger of being cut off from Heligoland and set out for that place at full power. About 8:50 the outlines of two

[144] This text extract and map is from the U.S. Office of Naval Intelligence translation of Krieg zur See, Der in der Krieg Nordsee.
[145] V-187 is the Flotilla leader at this time, see map at end of this Appendix for flotilla vessel placement.
[146] German naval charts subdivided the North Sea into quadrants being the a letter of the Greek alphabet. Within the Greek referenced quadrant was a number numbered grid subdivision.

four-stack cruisers in column were made out on the port beam at 5450 to 6500 yards range. In view of the radio reports, these could only be considered hostile. V-188 at once turned off two point to starboard but she had already been seen as the leading cruiser challenged her by searchlight which was answered with convenient letters. Immediately thereafter the Captain sighted two more cruisers apparently of the same type further forward, about two points, forward of port beam. He turned off two more points to starboard to avoid them. In doing so he now sighted the adjacent boat on the right, V-190, Captain Lieutenant Waitz, which a little later formed a column on V-188. After another exchange of false recognition signals, the enemy was no longer deceived, and at 9:07 opened fire with well laced salvos at 6550 to 8750 yards. At 9:12 an additional two British cruisers were sighted on port beam and V-190 reported by radio: "Six enemy cruisers in 132 epsilon." The last two ships sighted participated in firing for only a short time and then turned off to the westward apparently looking for other boats., while the other cruisers chased in the direction of the Jade. At this time the forward main feeding pump on V-190 broke down, so that the boat could only make 18 knots for the time being. At first during the pursuit only two cruisers continued to fire without support, however, making hits, while the other two apparently were firing at targets on their port side. Not until 9:35 did these also take a hand in the action at about 4500 yards range and right afterwards for no apparent reason they turned off to the westward. Shortly before three destroyers were sighted three point on the port bow. At the same time appeared to southward the remaining boats of the Half-Flotilla, B-197, V-191- and V-189, Captains, Lieutenants Richard Beitzen, Froelich and Metger. Upon orders of Commander of Torpedoboats, these had taken course for Heligoland at about 9:10 and already joined up when at 9:25 they sighted the other two boats in action with hostile cruisers, of these, however, only two could be made out from the leader. Believing that he was cut off from Heligoland, the Half-Flotilla Commander on V-191, Lt-Commander Tegmeyer, at once took course southeast by east for the Jade and at 9:29 opened fire on the leading cruiser at a range of about 4350 to 6550 yards, so as to relieve the hard pressed boat above, V-190. Hits could not be observed on account of the hazy atmosphere. At the same time th radio report was made: "Two enemy cruisers 133 epsilon chasing Half-Flotilla I."

The forward hostile cruiser answered at once with a spirited salvo-firing of 3.9 and 5.9 inch guns[147]; but the shots of the smaller caliber did not detonate. Neither did the opponent keep salvos in the air, as we usually do, but waited each time for the impact, so that by frequent turns of 1 to 2 points following the flash of the shots we were able to evade the impacts. During this time visibility changed so that even during the action the cruisers could not always be seen equally well. At 9:33 they turned hard to westward and then could be made out as Town class cruisers

[147] Actually 4" and 6" respectively.

with four stacks and two masts. Scarcely had this movement of the enemy been reported by radio when two more hostile cruisers were sighted, which took V-188 and V-190 under a spirited fire. To relieve the latter boat, V-191 again opened fire on the leading cruiser, which then stood toward her new opponent and took the leader boat under a spirited, and for the most part, covering salvo-fire. In this instant a submerged submarine was suddenly sighted ahead and a second one only 450 yards to starboard; both disappeared again beneath the surface after two shots had been fired at the former at 100 yards range.

Shortly thereafter , at 9:20, the cruisers broke off the engagement for no apparent reason and turned to the westward. This, as well as the sighting of the submarine, was reported to the High Command by radio.

Meanwhile, V-188 and V-190, so as not to be cut off from the Jade, had turned towards the destroyers. A short passing engagement took place with them at 7650 to 5450 yards range but apparently neither side scoring hits. Once again it was observed that the majority of the British 4" shells did not detonate. After 15 minutes the destroyers, apparently ships of the Acheron class, turned away, but the German boats did not follow them because the enemy cruisers were still in the immediate vicinity and did not permit the other boats of the Half-Flotilla to close in support. After the hostile force had disappeared from sight, all the boats joined up at 10 o'clock and set course for Heligoland to await further orders upon arrival there under the protection of the fortifications.

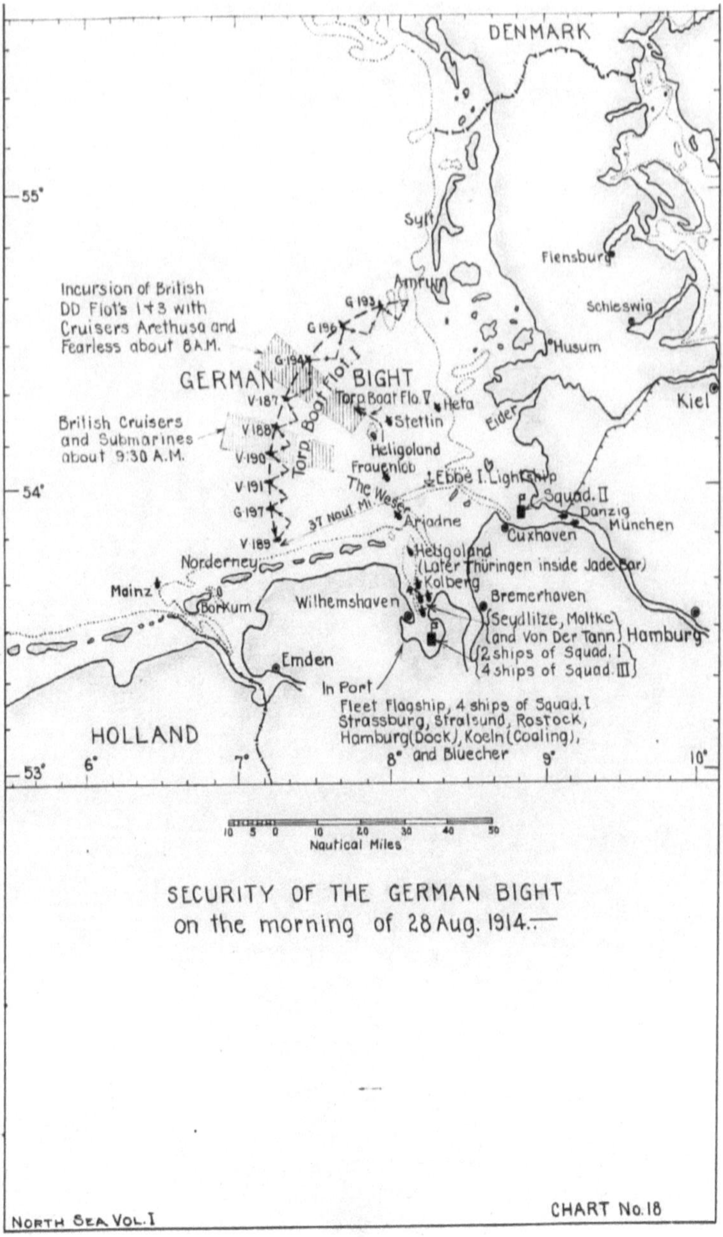
DENMARK
Sylt
Amrum
Flensburg
Schleswig
Husum
Kiel
Incursion of British DD Flot's 1+3 with Cruisers Arethusa and Fearless about 8 A.M.
GERMAN BIGHT
Torp. Boat Flot. I
Torp Boat Flo. V
Heta
Stettin
Eider
British Cruisers and Submarines about 9:30 A.M.
Heligoland
Frauenlob
Ebbe I. Lightship
The Weser
37 Naut. Mi.
Ariadne
Squad. II
Danzig
München
Cuxhaven
Norderney
Heligoland (Later Thüringen inside Jade Bar)
Kolberg
Mainz
Borkum
Wilhemshaven
Bremerhaven
Seydlitze, Moltke and Von Der Tann
Hamburg
2 ships of Squad. I
4 ships of Squad. III
Emden
In Port
Fleet Flagship, 4 ships of Squad. I Strassburg, Stralsund, Rostock, Hamburg (Dock), Koeln (Coaling), and Bluecher
HOLLAND
Nautical Miles
SECURITY OF THE GERMAN BIGHT on the morning of 28 Aug. 1914.
NORTH SEA VOL. I
CHART No. 18

Appendix C
I Torpedoboat Half-Flotilla movements following Whitby/Hartelpool Bombardment Dec 16, 1914.[148]

Overleaf: Map of relative maneuvers of German and British light forces following the bombardment.
The I Half-Flotilla was lead by the cruiser Graudnez. Plot commences at 12:30 and follows the efforts of the Germans to break eastward to the Heligoland Bight and the intercepting British cruisers referred to in the text and notes.

[148] Map is from the U.S. Office of Naval Intelligence translation of Krieg zur See, Der in der Krieg Nordsee

V188

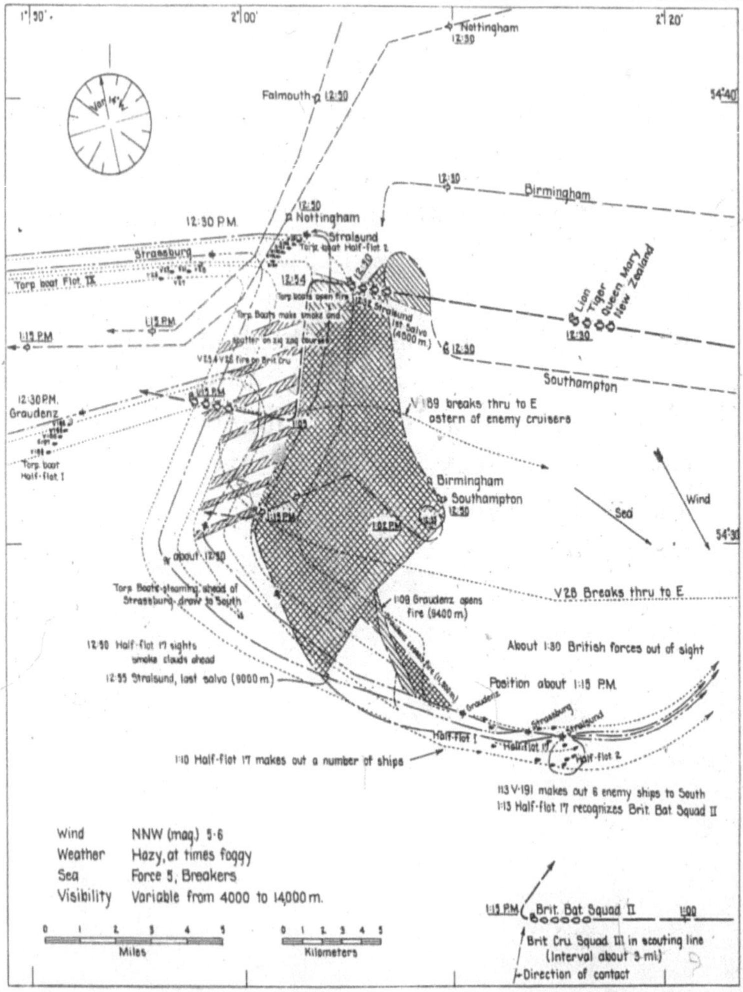

ACTION AND BREAK THROUGH OF THE CRUISERS STRALSUND, GRAUDENZ, STRASSBURG, AND TORP BOAT FLOTS I and IX, 16 Dec. 1914

The War in the North Sea Vol. III — Chart 12

Appendix D
German Official History Account of the Sinking of V188[149]

On the 25th of July, the work on this site[150] was resumed, in particular by the I Minesweeping Division.[151] In the process, the north-east and south-west of the barrier were finally found in 074[152] and 073[153] and marked by a red Elb and Ems buoys respectively. At the same time as the I Minesweeping Division, the I Torpedoboat Flotilla, Lt. Konrad Albrecht, had left the Ems and had taken over the security of the minesweeping division. To this end, about twenty miles west of the island, an outpost line of about thirty miles extended, in which the five boats of the half-flotilla drove back and forth individually at high speed on zigzag courses. At midday a hostile submarine was sighted at noon, the minesweeping was discontinued and the submarine hunted in vain. When the torpedo boats steamed back and forth on the outpost line on the following day (26th of July),

[149] pp.209-210 Der Krieg zur See, Der Krieg in der Nordsee vol.4, Otto Groos. Berlin, 1924.
[150] Reference is to a discovered British laid minefield.
[151] Minesweeping had been interrupted due to the activities of British submarines in Heligoland Bight. Note Appendix F indicating British submarine sightings by German forces in the Bight during July 1915. The sinking of V188 culminated a month of near constant reports of sightings with British submarines resulting in the laying of deeper and more elaborate German defensive minefields to obstruct British submarines.
[152] Map grid reference
[153] Ibid

one could suddenly see from the leader boat "V 186" at 1.36 sea miles distance a white cloud of steam ascend over "V 188", Lieutenant Callisen. Being signaled all the boats approached at high speed towards "V 188" which, badly damaged by a torpedo hit and unable to maneuver, fired with its forward gun on a submerged submarine. A few minutes later V186, Reserve Lieutenant Boy Ed, was also fired upon from the submarine at about 300m. But the commander was still able to evade the torpedo by using "full speed ahead" and hard rudder maneuvering, while a torpedo of his own, with an 8 meter depth setting aimed at the clearly recognized swirl of the submarine, failed. The periscope came again in sight, but also a ramming failed.

Shortly thereafter V188 went down, firing to the end at the sight of the submarine. It succeeded, although always threatened by new submarine attacks, to save its crew in dinghies; but watch officers, Reserve Lieutenant Gulecke, Lieutenant Lamm, and Half-Flotilla Naval Engineer Otto Karl Meyer, and two men were missing and probably lost their lives as a result of the explosion. The Half-Flotilla and the Minesweeping Division were recalled immediately to the Ems on the announcement of these events. However, on the following day, as they were still standing in the work area, the 4 boats of the I Half-Flotilla once again managed to hunt down the enemy submarine, and finally bombarded a suspicious oil spot north of the sinking of V188 of about 100 meters long with 8 explosive cartridges.[154] Depth charges, which later turned out to be awesome weapons, the torpedo boats had not yet on board at that time, compared to English anti-submarine defenses who had already been threatened by the new weapon[155] were better trained at the time.

[154] There was no submarine present at this location. Note Appendix G, Log of H.M. Submarine E16 which indicates no interaction with German Forces on this date.
[155] The submarine is the new weapon.

V188

Appendix E
Reported enemy submarines in the German Bight in July 1915th[156]

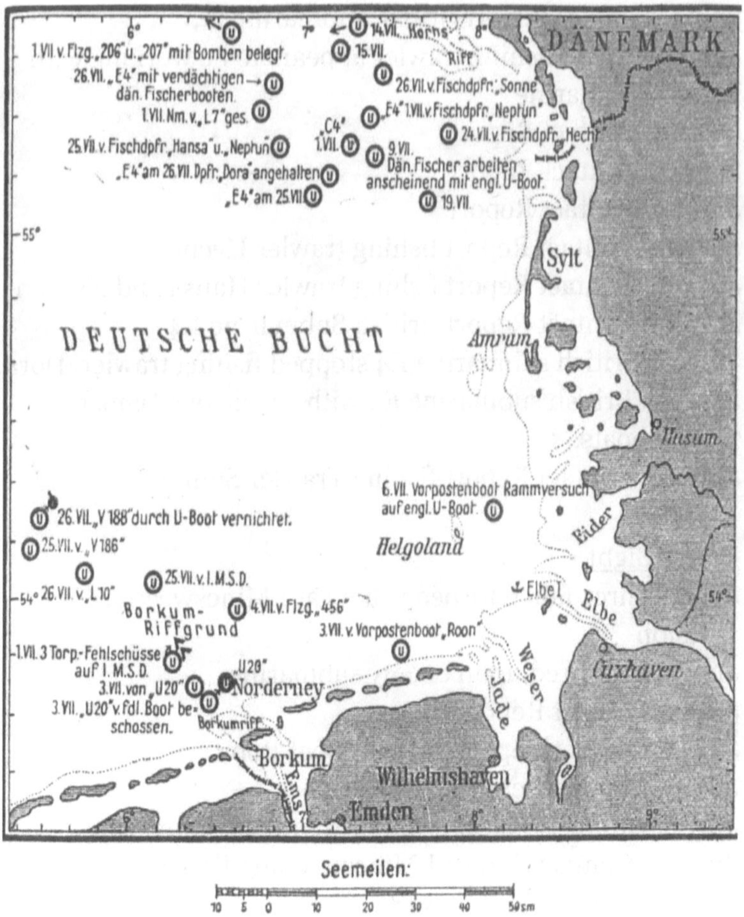

Gemeldete feindliche U-Boote in der Deutschen Bucht im Juli 1915.

Overleaf. Contacts detailed in map:

[156] p.211 Der Krieg zur See, Der Krieg in der Nordsee vol.4, Otto Groos. Berlin, 1924.

Upper Bight

July 1. Aircraft nos. 206 and 207 bomb
July 1. Airship L7 fire upon
July 1. Contact Report British submarine E4 by fishing trawler Neptun
July 1. Contact Report British submarine C4
July 9. Danish fishing trawler appears to be working with British submarine
July 14. Contact Report
July 15. Contact Report
July 19. Contact Report
July 24. Contact Report fishing trawler Hecht
July 25. Contact Report fishing trawler Hansa and Neptun
July 25. Contact Report British Submarine E4
July 26 British submarine E4 stopped fishing trawler Dora
July 26. British submarine E4 with suspicious Danish fishing boats.
July 26. Contact Report fishing Trawler Sonne.

Lower Bight

July 1. Three failed torpedo shots at I Minesweeping Division
July 3. U20 fired upon enemy submarine.
July 3. Contact Report U20
July 3. Contact Report Outpost Boat Roon
July 4. Contact Report Aircraft 456
July 6. Outpost Boat ramming attempt at submarine
July 25. Contact Report I Minesweeping Division
July 25. Contact Report V186
July 26. Contact Report Airship L70
July 26. V188 destroyed by submarine

Appendix F
Callisen's report on the Loss of V188[157]

Occurrence

The torpedo boat half-flotilla was to provide westward security for a minesweeper division on 26 July 1915. For the purpose an outpost line was designed pointing from the N-S direction. The position for V188 as northern wing-boat was square epsilon 051 according to the half-flotilla order. In square center epsilon 066 V188 was dismissed by the leader boat. I steered to the bottom left corner of my allotted square. On course before arriving there observed in square epsilon 037 a trawler in sight, which I then held towards.

After completing the investigation I steered for 1hr 20 m to my position on a Northeasterly course at 20 knots. The trawler very soon disappeared in a thunderstorm. After about 7 minutes reported a lookout at the front mast reported a submarine 6 points to starboard. Unfortunately a range[158] report did not follow. I personally was on the port side of the bridge, therefore, I did not initially see the submarine. I instinctively sensed: "It", I quickly commanded "Hard starboard, Starboard engine half speed astern" to possibly position the bow on the submarine. As I

---

[157] 1st Torpedoboat flotilla KTB. Roll 75 PG 62072 NARA T1022 Captured Records of the German Navy

[158] For clarification, range to the submarine.

sprang to the other side of the bridge, I already saw the bubble track coming at about 100m, intersecting at 90° to my old course direction. The periscope was at no more than 150m distant in the midst of a large circle, from which originated the shot. [159] It appeared to me evading the torpedo was immediately ruled out in view of experiences in peace shooting exercises. The possible impact lay far to the stern and to make the angle of impact as obtuse as I possible, I immediately went full speed ahead on the port engine. The torpedo hit the middle of Division III and instantly disabled the boat.

The detonation took place very late, when the faint track was already at the boat, so I held the view that the boat was undershot. The blast cloud was slightly higher than the mast, deep black, although a coal bunker was not hit. The boat was torn apart in Division III; the two ship parts only connected together by some pieces of the port board sides and apparently the port shaft. The bulkheads to Division II and IV were strained, in Division IV the condenser and many steam pipes were damaged, so that steam danger was present. As for the rest of the boat equipment the shock immediately put the spotlight out of operation, and the din is not very violent (like that after a strong collision). The gas pressure on the bridge was not felt by the crew who were on deck, only one man, who standing behind the tower, was knocked down, but not injured. The ship's debris flew

---

[159] The reference is to the disturbed water following a torpedo discharge.

mostly over to port because the wind came from starboard aft (strength) 4.[160]

The boat buckled immediately and sank with Division III so deeply under that the upper deck was level with the water surface. The ship's stern floated as well, however with about 30° of the starboard side canted up. The bulkhead between Division IV and V held, so that the boat was about 15 minutes at the same depth.

With the measures now taken appeared 2 obstacles. One the lack of all officers and off watch petty officers below deck in the aft had no connection with those forward, so I had to rely on the support of the few deck petty officers in the bow. Further, the noise of escaping steam was so great, that it was impossible, to get through by voice only up on deck. I therefore had all commands dispatched forward by runners.

To me it was initially important to save the boat and prevent the submarine a second shot. Therefore I had the bulkhead braced between Division IV and V under the direction of senior machinists and the forward gun, under the direction of the coxswain, to bombard the periscope in sight on the starboard side astern. With the same gun I had in between shooting, shoot a few shells at high elevation in the direction of V186, which was about 6 miles distant to draw attention to myself. Due to no.1[161] I had then to make clear to the F.T. seaman "how to" deploy the reserve antenna and

---

160 Reference to Beaufort Wind scale

161 No. 1 being the absence of officers and petty officers forward of the point of torpedo impact.

finally the auxiliary helmsman to launch the dinghy in order to bring the crew from off the stern to the foredeck.

After these preparations I had, although men at sea constantly wear inflatable life jackets, to "ready the lifejackets" to the extent available also using the safe cork life jackets.

After about 15 minutes V186 approached. I had no quick means to signal an alert to the submarine danger, so I continuously fired at the spot where the submarine was, later marking from the yard of the F.T. mast a semaphore signal.

V186 passed by at high speed, so that the wake made the boat roll. A moment later, I received a message that the bulkhead would no longer hold between Division IV and V. I therefore suspect the leakage was due to the breaking movement at the buckle spot rupturing the bulkhead. In this I was confirmed by the fact that the portside bulkhead broke, also on the side where the two ship parts hung together. The boat now listed to portside and thus it was clearly doomed. Just as the boat was heavily over to port a submarine was reported ahead with her periscope raised out high. I immediately had the cannon shoot at it, but could not prevent it firing a torpedo at V186. The boat was already so heavily [listing] over that the cannon could no longer pivot, but had to be turned around by force.

As capsizing was imminent, I had the middle deck crew start to get the dinghy alongside, and place the injured inside. Also maps with mine registrations, which were previously tied together, and the bridge signal book were brought in. The signal book key from the F.T. cabin I placed with myself.

After a 3 times Hurrah to his Majesty the Emperor, which was enthusiastically received, I ordered: "All men over board"! the boat was already so heavily [listing] over that I was no longer able as the last to jump down from the bridge, but I could go down only on deck.

The boat steeply straightened up at a high height then descended and sank downward upright.

The crew was taken off by the cutter and dinghy of V186 and the dinghy of V188.

Casualties

All persons in Division III are not seen again after the detonation, undoubtedly killed immediately. I myself operated the alarm bell on the bridge shortly before the detonation; but it was probably not possible to get out of the Division. The corpses remained in the boat, namely:

1. Reserve Senior Lieutenant[162] Guleke

2. Lieutenant.[163] Lamm,

2. Marine Chief Engineer[164] Meyer (Otto Karl)

---

[162] Reserve Oberleutnant zur See in German. Full name Georg Guleke

[163] Leutnant zur See in German. Full name Karl Lamm

3. Torpedo Seaman Baumgartel[165] (Officer's steward)

4. Torpedo Chief Stoker Felsch[166] (Orderly)

All the others are saved. Injured were several lightly due to splinters, 1 petty officer and 1 man (both from Division IV) due to scalding.[167] These two are in the infirmary in Emden.

Experience

1. Although the torpedo detonated immediately adjacent to the ammunition compartment and under the starboard aft rotating tube (pistol[168] at safety position), there was no further detonation.

2. In my opinion there were 2 submarines there at 500-700 meters that maneuvered simultaneously. I myself have seen two periscopes in quick succession, so I do not think that the distance between the two locations could be covered so quickly. In my opinion the shot at V186 comes from another submarine.[169]

3. Particularly distracting seemed, as mentioned above, the roar of escaping steam. It was not possible to use the siren as signaling medium because the power supply was then disabled at the Bridge.

---

[164] Engineers were differentiated from sea officers. Hence the Marine indicator.

[165] Ober Matrose in German. Full name Oskar Baumgartel

[166] Ober Heizer in German. Full name Konrad Felsch

[167] Division IV the engine/boiler room of the Torpedoboat hence the scalding injuries due to ruptured steam lines.

[168] The torpedo pistol is the firing mechanism within the warhead. Adjusted via exterior means.

[169] There was only one submarine present, E.16.

4. The boat was instantly, and I suppose that is usually the case, deprived of the most useful distance signaling methods (the spotlight and the F.T.[170]), so that I was dependent on the attention of all neighboring boats.

Crew Behavior

The attitude of the crew was superior above all praise. Not even a single one shrank for an instant from their role. I had the impression that everyone thought only to save the boat. The dinghy remained lashed until my command to uncover, finally no one jumped overboard before my command. To me it is impossible to name every individual who has distinguished themselves. I cite only as an example the work of senior machinists, further including a signalman, who carried the heavy signal book from the bridge and wondered if he should take it, the helmsman and gun crew that shot at the periscope right to the end. When I jumped last into the water came an expression of the crew's spirit. As rang out from the men a loud 3 times hurrah for V188 and the captain. I would not like to leave this unmentioned, if it sounds too strange, to properly grant my crew their right. As the above faded sounded again a second 3 times hurrah resounded from the crew of V186.

Finally I would like my report not to end without mentioning the behavior of Sea Cadet[171] Ense of V186.

---

[170] F.T.=Funken Telegraph = wireless transmitter

[171] Fähnrich zur See in German.

He went alone in V186's dinghy among the swimmers, picked up some and then jumped into the water, to provide us space on the boat.

sgd.

Callisen Lieutenant[172]

[172] Kapitänleutnant in German.

Appendix G
E16's Patrol Report on the sinking of V188[173]

24th July,
9.0 a.m. Left Yarmouth.
5.0 p.m. Position: 53°-6' N., 3°-50' E. (Yarmouth 120miles)
Released two pigeons.

25th July,
0.45 a.m. Dived to bottom in 14 fathoms, N.39 ., 17 miles
from Borkum Light Tower.
4.30 a.m. Rose to surface Schiermonnikoog Towers in sight.
5.0 a.m. Dived. Shaped course for Huibert Gat.
5.20 a.m. Through periscope, sighted enemy Submarine
come to surface about 1-1/2 miles to Westward. Altered course to attack. The Submarine appeared to have the line of the upper deck largely curved fore and aft, and there was no casing. Enemy started up engine and proceeded to the Westward. Her speed did not exceed 10 knots.
6.36 a.m. Rose to surface and chased at full speed.

---

173 ADM 137/2068 Commodore (S) war records, Volume II. Reports of proceedings of submarines attached to HMS Maidstone. pp.447-450. Abbreviated version in ADM 173/1234 Log of E16.

6.45 a.m. Enemy dived, when about 3 miles distant. Continued to Westward at full speed to get ahead of him, but at

6.57 a.m. Was forced to dive by appearance of a seaplane.

7.15 a.m. Zeppelin to Southward.
Worked out to Northward.

8.30 a.m. Five T.B.D.'s appeared, steaming fast on various courses till about 10.0 a.m.

11.30a.m. Rose to surface, but after 10 minutes was forced under by Zeppelin.

Noon. Position: 53°48’ N., 6°0’ E.

Noon. While at 65 feet, bomb exploded vicinity of Boat.
Went to bottom for one hour.

1.27 p.m. Seaplane to South-Eastward.

2.15 p.m. Position: N.W. by N1/2N 3 miles from position of Borkum Riff Light Vessel.
Diving at 22 feet, 2 knots. Course 60°
Boat suddenly becomes heavy by the bow, and gradually sank with a big and increasing inclination down. Tried to correct trim but it needed a very large proportion blown from 1,2,3, and 4 Externals before boat was buoyant.
Two small bombs exploded near Boat.
Sank to bottom in 15 fathoms.
Went astern, only effect being to lift stern up to big inclination (more than 10°, off gauge).

2.40p.m. Blew 1,2, 3 and 4 Externals and rose to surface, still with large inclination down;

opened Conning Tower lid, to discover Zeppelin overhead, having just passed over boat; saw nothing foul forward.
Sank to the bottom.
Went astern and ahead for a few moments each; then blew 1,2,3, and 4 enough to lift Boat a few feet off bottom, and went astern at 4 knots, diving the boat horizontally.

3.15 p.m. Boat's nose suddenly flew up and Boat rushed towards surface. Checked it by vents, and found the Boat apparently free of the obstruction with her more or less normal trim.
Dived out to the Westward at 60 feet.

3.40 p.m. Rose to 22 feet. Nothing in sight.

4.0 p.m. Bomb exploded close to Boat. Returned to 60 feet, and remained diving till dark, fearing I might be towing a mark buoy.
While foul of the obstruction, the boat's position must have been very visible owing to the large amount of blowing, pumping and venting of tanks necessary.

8.0 p.m. Rose to surface, course 250°, charging batteries.

<u>26th July,</u>

1.0 a.m. Position 53°-50'N., 4°-20' E.
Endeavoured to get in W/T touch by means of kite, but after many attempts at flying them, both kites pitched very badly and were eventually lost overboard.
While doing this, a trawler of the same build

as German Patrol trawlers passed 5 or 6 miles to the Eastward, steaming South very fast.

4.0 a.m. Released 4 pigeons, with following message:- "To "Maidstone". Sunday afternoon fouled N.W. by N."3 miles from Borkum Riff Light Vessel, submerged "obstruction apparently consisting of heavily moored "net, watched by enemy's aircraft. I am now clear. "to station, leaving minefield to Southward".

4.38 a.m.Proceeded 12 knots. Course 51°.

5 to 6a.m.Diving to investigate and then avoid trawler.

6.15 a.m. Rose and proceeded.

8.45 a.m. Dived for Zeppelin to South-Westward, which was in Sight more or less continuously till 11.20 a.m.

11.20a.m. Sighted smoke and masts appearing from South- Eastward. Course as requisite to investigate.

Position 54°-11' N., 5°-23 E.

Three German T.B.D.s appeared, steaming independently at varying speeds on various courses.

12.36p.m.Fired port bow tube at T.B.D. at 500 to 600 yards range. Allowed deflection for 15 knots, hit about ¼ of her length from aft.

The other two T.B.D.'s approached at full speed from different directions.

The vessel hit appearing certain to sink, her stern being practically blown off, attempted to torpedo one of the others; fired starboard bow torpedo at 12.55 p.m. and stern

torpedo at 1.8 p.m., both missing In the first case, T.B.D.'s telegraphs must have been rung to full astern the instant before I fired; in the second case the torpedo was avoided.
Damaged T.B.D. had sunk by 1.5 p.m., a large proportion of her crew were taken off by other T.B.D.'s, which apparently picked up boats in intervals between attempting to ram "E.16". One T.B.D. then disappeared to Eastward, and remaining one, with a new arrival, apparently commenced sweeping at about 20 knots.
Dived out to North-Eastward.
Was kept diving until 7.0 p.m., a Zeppelin being almost continuously in sight until then.

7.0 p.m. Rose and proceeded in to station, leaving Borkum Flat minefield to Southward.

27th July,

0.15 a.m. Dived to bottom, position 54°-4'N., 6°-55' E.

5.0 a.m. Rose and proceeded in towards Norderney on surface.

6.30 a.m.Dived.

8.20 a.m.Surface. Seaplane to South-Westward.Dived. Diving to N. and N.W. of Norderney Island, rising to surface at intervals. Saw one merchant vessel passing close inshore, steering East.

3 to 9p.mPatrolling on surface between Borkum and Norderney, very heavy rain and thunder prevented appearance of aircraft.

9.0 p.m. Dived to bottom, 53°-55' N., 6°-57' E.

28th July,

Wind S.W., 5-7, Sea rough. Too rough for aircraft.

Patrolling on surface and submerged.

Saw a tug come out from Eastward past Norderney, and later pilot a merchant steamer in to Eastward, and in afternoon got within 3,000 yards of steamer proceeding from Eastward past Norderney; about 3,000 to 4,000 tons, painted dark colour, no marking or colours

9.0p.m. Dived to bottom. Position 55°-55' N., 6°-58'E.

29th July,

4.30 a.m. Rose off bottom.

Wind West, 6-7. Sea rough. As yesterday.

Saw three steamers pass close in shore.

9.0 p.m. Dived to bottom. Position 53°-57'N., 6°-50'E.

30th July,

4.30 a.m. Rose. Wind W.N.W., 7, Rough.

Patrolling mostly on surface.

Saw two steamers and one T.B.D. pass close in shore.

4.30 p.m. Worked out to Northward at 6 knots to pass Round Borkum Flat minefield, and then or Harwich.

Norderney Light Tower has been shifted to the extreme eastern end of the island.

All vessels seen in the area between Norderney and Borkum were on, or inshore of, the 5 fathom line; this route is apparently not swept, as no trawlers were seen.

C.P. Talbot[174]
Commander

[174] Cecil P. Talbot,(1884-1970), earned the DSO for sinking V188, He also sank the German submarine U6 In September 1915. Surviving the war he retired as a Rear-Admiral.

Illustrations from the Original Work

My Boat

Badge: two crosses on the mast

The missing V1 before the digits is a common service abbreviation

Overleaf: V188 in storm. Note the steering rudder visible beneath the bow unique to some boats of this class.

V188

Smoking the Boilers

Among ice in the foreboat's wake

Iced Over

Ten waves roll over every minute

The Half-Flotilla at Sea

Against the Sea

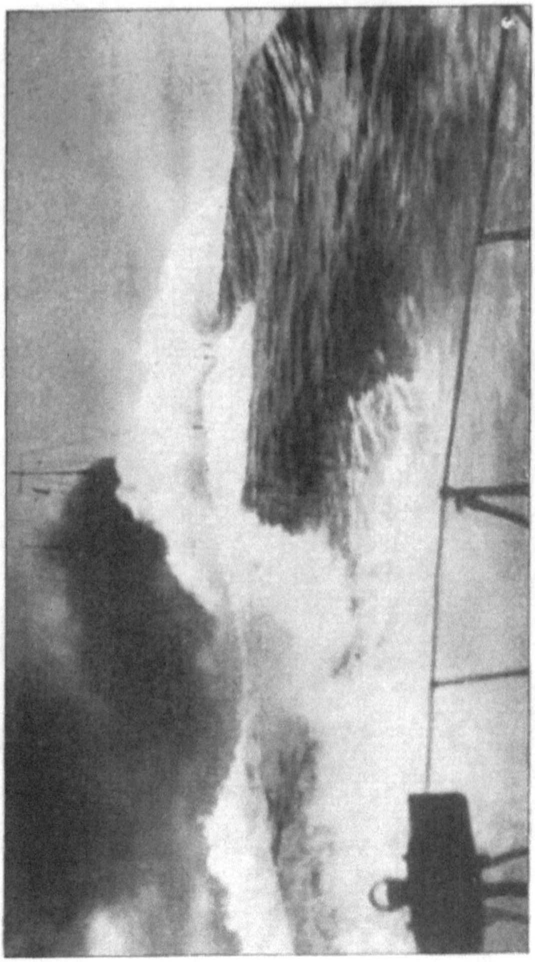

Loading a torpedo

Torpedo shoot: Readying

Torpedo shoot: Torpedo leaves the tube

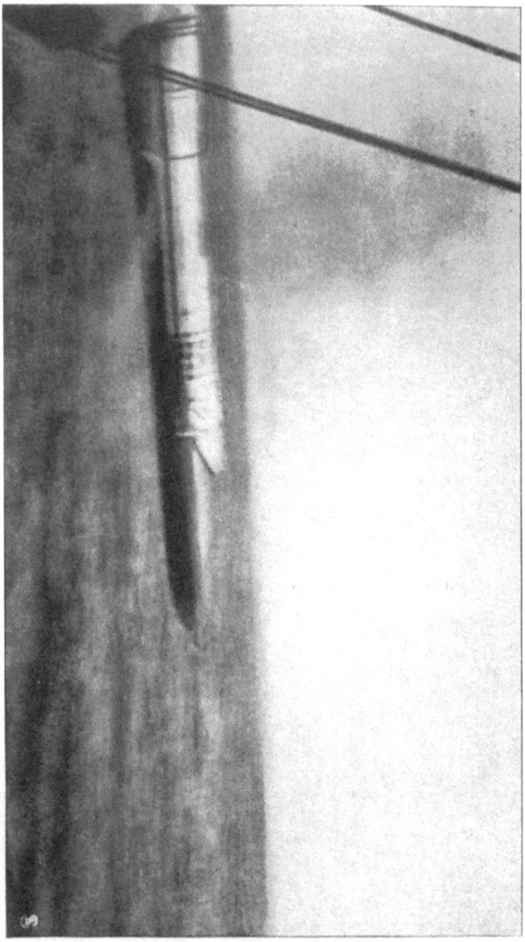

Torpedo shoot: Torpedo in the air

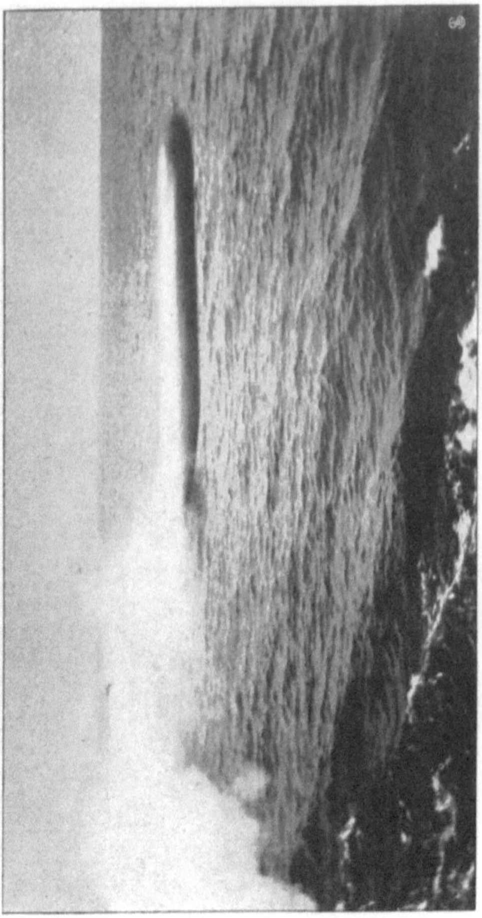

Torpedo shoot: Torpedo hits the water

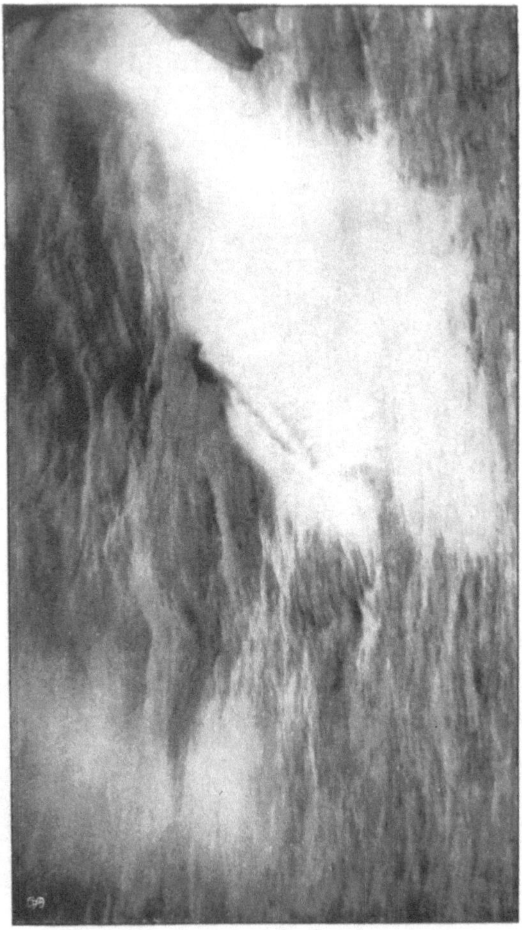

Eliminating a mine.

Lunch on the anchor capstan

Torpedo boats in the occupied port of Libau

Made in United States
Troutdale, OR
11/03/2026

56852532R00100